The Aesthetics of Scientific Data Representation

How can cartoon images aid in understanding bacterial biological processes? What prompts physicists to blur their images before showing them to biologists? Considering that the astronomer's data consists solely of invisible, electric impulses, what is the difference between representing outer space as images, graphs, or sound? How does a work of contemporary art differ from a scientific image if we cannot visually distinguish between the two? How does aesthetics, art, and design influence scientific visualization and vice versa? This volume asks critically important questions about scientific data representation and provides significant insights to a field that is interdisciplinary in its very core. The authors investigate scientific data representation through the joint optics of the humanities and natural sciences. The volume particularly appeals to scholars in visual and aesthetic studies, data visualization, scientific illustration, experience culture, information design, and science communication.

Lotte Philipsen is an Associate Professor with the School of Communication and Culture, Aarhus University in Denmark and former Fellow with the Aarhus Institute of Advanced Studies, Denmark.

Rikke Schmidt Kjærgaard is executive director of Science Club, Denmark, CEO of Graphicure, and former Fellow with the Aarhus Institute of Advanced Studies.

Routledge Advances in Art and Visual Studies

A full list of titles in this series is available at: https://www.routledge.com/Routledge-Advances-in-Art-and-Visual-Studies/book-series/RAVS

The Aesthetics of Scientific Data Representation

More than Pretty Pictures

Edited by Lotte Philipsen and
Rikke Schmidt Kjærgaard

Routledge
Taylor & Francis Group

LONDON AND NEW YORK

First published 2018 by Routledge

2 Park Square, Milton Park, Abingdon, Oxfordshire OX14 4RN

52 Vanderbilt Avenue, New York, NY 10017

Routledge is an imprint of the Taylor & Francis Group, an informa business

First issued in paperback 2019

Library of Congress Cataloging-in-Publication Data
Names: Philipsen, Lotte, editor. | Kjærgaard, Rikke Schmidt, editor.
Title: The aesthetics of scientific data representation: more than pretty pictures / edited by Lotte Philipsen and Rikke Schmidt Kjærgaard.
Description: New York: Routledge, 2017. |
Series: Routledge advances in art and visual studies |
Includes bibliographical references and index.
Identifiers: LCCN 2017018907 | ISBN 9781138679375 (hbk: alk. paper) | ISBN 9781315563411 (ebk)
Subjects: LCSH: Natural computation—Graphic methods. |
Visual communication—Data processing.
Classification: LCC Q222 .A37 2017 | DDC 500—dc23
LC record available at https://lccn.loc.gov/2017018907

ISBN: 978-1-138-67937-5 (hbk)
ISBN: 978-0-367-87869-6 (pbk)

Typeset in Sabon
by codeMantra

To all cross-disciplinary researchers who dare to step outside the confinements of their core discipline.

Contents

Figures and Color Plates

Figures

Color Plates

Preface

Lotte Philipsen and Rikke Schmidt Kjærgaard

The Aesthetics of Scientific Data Representation: More than Pretty Pictures (Routledge Advances in Art and Visual Studies series) explores scientific data representation across boundaries and scopes of the humanities and the natural sciences, discussing diagrams that map complex chemical relations, scanning microscopy providing images of nanoscale structures, sonic output testifying to tectonic movements, and colorful images of galaxies in outer space where no light is ever present (to mention just a few examples from this volume).

Representation, scientific data representation, too, is inevitably a matter of aesthetics since all representations are created and shaped under human influence. Even a graph on a monitor have at least shape, color, position, and size, and however advanced devices we may use in the process, these representational features are culturally and technically constructed by humans.

Natural scientists constantly make use of and depend on aesthetic representations in their research but tend to consider such representations as neutral transmitters of data. On the other hand, researchers in aesthetic theory have in-depth knowledge on image representation, aesthetics and epistemology, but tend to apply this knowledge to analyses of art or popular culture, and ignore the ever-important role data representation plays within the natural sciences.

The aesthetics of scientific data representation is a cross-disciplinary matter – accordingly, the intended readership of this volume includes researchers and students from the natural sciences as well as from the humanities. Despite the cross-disciplinary nature of this field, when navigating it in practice, most of us tend to follow familiar routes within the established boundaries of our own discipline. Terminologies that most of us take for granted, like "artists", do not mean the same within different disciplines. We realize this only when forced to explicate whether we mean "artists", "designers", "illustrators", "data visualization experts", or something else. Similarly, different scientific domains have different understandings of "data", "representation", "aesthetics", etc. This volume addresses these challenges through cross-disciplinary, comprehensive understanding of the epistemic, aesthetic, and communicative mechanisms and implications of contemporary scientific data representation.

Existing interdisciplinary resources on scientific data representation have so far mostly been dominated by an art-science focus and historical accounts. Edward Tufte's works on graphic design for data representation are among the most recognized examples of guides on how to represent data in ways that combine clarity with visual attractiveness (Tufte, 2001, 2003, 2006). However, they convey an ideal of prettiness and clarity in the production of scientific visual representations without acknowledging the aesthetics that lie beyond the pretty. Other books touching upon art and the conception of data representation are Katy Börner's huge coffee-table books on mapping. However, Börner does not explicitly address the concept of aesthetics in data visualization (Börner, 2010, 2015). Peter Galison and Caroline Jones' edited *Picturing Science, Producing Art* accounts for relations between art and science, too, but its chapters explore science-art relations from a wide variety of different theoretical and historical perspectives (Jones & Galison, 1998).

Over the years, art historian James Elkins has contributed with valuable research on scientific images. Elkins's book *The Domain of Images* (1999) offers an intriguing take on scientific images, although from an art historian's point of view; Elkins's *Six Stories from the End of Representation* and his edited volume *Visual Practices Across the University* include viewpoints of both humanists and natural scientists and offer insights into the roles images play in a wide variety of disciplines (Elkins, 2007, 2008). A similar approach is seen in the volume edited by Catelijne Coopmans et al. (2014), *Representation in Scientific Practice Revisited*, which focuses on contemporary practices of scientific representation through a variety of interdisciplinary investigations.

An important contribution to understanding the cultural constructions of scientific imagery is Peter Galison and Lorraine Daston's book on the history of *Objectivity* (2007), which has a historical rather than contemporary focus, and a similar theme is investigated in Lisa Gitelman's edited volume *"Raw Data" Is an Oxymoron* (2013).

As such, the overall theme of the current volume belongs to a well-established field of investigations in scientific data representation. However, this volume focuses exclusively on contemporary practices of scientific data representation – historical references are sparse and serve only as a means to better comprehend current practices. It offers in-depth aesthetic analyses that reaches beyond the average how-to-guide on data visualization and, most importantly, does not reduce the question of aesthetics to a matter of pretty pictures. It does not, however, systematically cover data representation from all scientific disciplines. Some chapters focus exclusively on the aesthetics of data representation within one discipline, or even one specific example, while other chapters investigate a specific dimension of data representation across a number of disciplines. It is an edited volume authored by different experts across the natural sciences and the humanities that forms a whole in the sense that the chapters are informed by other chapters, offer numerous cross-references, and built upon each other in order to form a coherent progression throughout and across chapters.

The first chapter in this volume, "Visualizing the Bacterium *Streptococcus pneumoniae*" by Engholm et al., demonstrates the book's topic par excellence by providing a data representation that crosses the traditional borders of art and science: a hand-drawn watercolor painting of the bacterium *Streptococcus pneumoniae*. The painting is developed by a molecular biologist and very much inspired by other similar scientific and artistic paintings by molecular scientists and scientific illustrators. The chapter exemplifies the process of transforming data from the laboratory to the final painting through methods, specific choices, and considerations, producing a range of different images. The case presented in Chapter 1 implicitly prompts a number of theoretical questions on the relation between the icons and diagrams—and their different representational mechanisms—that are articulated and investigated in Chapter 2, "Diagrammatic and Iconic Imagery in Science" by Stjernfelt, who also introduces the role of beauty in data representations.

Chapters 3 and 4 elaborate on the diagrammatic and iconic dimensions respectively: Chapter 3 "Scientific Data Visualization: Aesthetic for Diagrammatic Clarity" by Krzywinski, discusses more broadly the ideas and key concepts of data design and data visualization. It creates a framework for data visualization based on examples and terms presented in previous Chapters 1 and 2, and in the author's own work focused on genome research. By looking at a wide range of examples, Chapter 3 considers the role of aesthetics in recent data visualization broadly. Chapter 4, "Plant(ing) Aesthetics between Science and Art" by Philipsen, maps different mechanisms of aesthetic judgment of taste in science, art, and critical design. The chapter combines its theoretical investigations closely with concrete analysis of three different representations of plants: a modified scanning electron microscopy, a piece of photographic art, and a (perhaps) fake sex consultancy service.

Some of the figures in the next chapter follow dogmas presented in Chapter 4, but the volume now moves from the molecular scale of the biosciences to the scale of the whole universe. Chapter 5, "Visualizing the Invisible Universe" by Hannestad, describes how astrophysicists look at and visualize outer space phenomena—things very big, very far away that, in most cases, have ceased to exist millions of years ago. The chapter portrays the rising complexity in visualization within astrophysics. From the challenges of taking photographs of outer space phenomena that emit light in wavelengths not detectable by the human eye and therefore in need of manipulation in the visualization process, to the methods of visualizing phenomena that are in their very essence invisible since they consist only in mathematical data. The morphology of data representation in relation to its production process is also the focus of Chapter 6, "The Epistemics of Data Representation: How to Transform Data into Knowledge", where Samuel discusses the epistemic roles of scientific data representation by demonstrating the gaps between phenomenon, data, and data representation. The chapter also demonstrates how scientists' perception of data representations depends on additional translational acts that compensate for the gaps—to the extent where physicists' data representation images need to be blurred for a biologist to comprehend them.

Chapter 7, "Sonification and Audification as Means of Representing Data" by Søndergaard and Vandsø, demonstrates how phenomena and data change according to the means of representation. Whereas the previous chapters almost entirely focused on visual data representation, Chapter 7 offers a much-needed insight into the domain of sonic data representations. Data of the chapter's examples originate from scientific domains (comets, cells, tectonic movements), and the chapter investigates how sonic representations are inevitably intertwined with artistic and mainstream cultural practices like Hollywood movies. Further investigation of the intertwinement between science and public audiences is conducted in Chapter 8, "Scientific Storytelling: Visualizing for Public Audiences" by Veldhuis, who explores considerations, negotiations, and discussions that go into the making of data representations for a non-specialist audience in museums, at festivals, and online. Whereas Chapter 8 discusses the concept of "selling science" in a broader perspective, to a broad audience, Chapter 9, "Communicating Science—Aesthetic Choices in Publishing" by Krause, focuses on the process of producing covers for the scientific journal *Nature* and takes the reader through the steps of understanding the research, understanding the "vernacular" or visual language of that particular field, and translating the message into visual form that is appropriate to the particular audience.

Rafner and Schmidt Kjærgaard extend ideas and concepts from general data representation to scientific animation and moving images in particular in Chapter 10, "Ideas in Action: Using Animation to Cut through Complexity". The chapter discusses the representational and aesthetic possibilities, advantages, and constraints that animation as a tool provides for data representation. Chapter 11, "Making Sense, Nonsense, and No Sense when Representing Audio-Visual Collections" by Vallø Madsen, investigates new methods for online visualization and interaction as a means of representing networks and meshworks of collections of audio-visual material. The entire volume is concluded in Chapter 12, "'Facts'—and Representational Acts" by Kyndrup, picking up from the trans-media discussions and key concepts in the previous chapters and exploring "What is representation?" and "What is *a* representation?" The chapter provides a basic, theoretical understanding of what constitutes and defines the notion of "representation", and it demonstrates its theoretical points by referring to selected specific cases and examples from previous chapters.

References

Börner, K. 2010. *Atlas of Science: Visualizing What We Know.* Cambridge: MIT Press.

Börner, K. 2015. *Atlas of Knowledge: Anyone Can Map.* Cambridge: The MIT Press.

Coopmans, C. et al. 2014. *Representation in Scientific Practice Revisited.* Cambridge: The MIT Press.

Elkins, J. 1999. *The Domain of Images.* Ithaca: Cornell University Press.

Elkins, J. 2007. *Visual Practices Across the University.* München: Wilhelm Fink Verlag.

Elkins, J. 2008. *Six Stories from the End of Representation: Images in Painting, Photography, Astronomy, Microscopy, Particle Physics, and Quantum Mechanics, 1980–2000.* Stanford: Stanford University Press.

Galison, P. & L. Daston. 2007. *Objectivity.* New York: Zone Books.

Gitelman, L. 2013. *"Raw data" Is an Oxymoron.* Cambridge: The MIT Press.

Jones, C. A. & P. Galison. 1998. *Picturing Science, Producing Art.* New York: Routledge.

Tufte, E. R. 2001. *The Visual Display of Quantitative Information.* Cheshire: Graphics Press.

Tufte, E. R. 2003. *Envisioning Information.* Cheshire: Graphics Press.

Tufte, E.R. 2006. *Beautiful Evidence.* Cheshire: Graphics Press.

Acknowledgments

We express our deepest gratitude to Aarhus Institute of Advanced Studies (AIAS) with Aarhus University, who helped us realize this book in numerous ways: by offering fellowships to researchers from a number of different academic disciplines, resulting in the chance meeting between a science communicator and an art historian; by generously providing the facilities, support, and time that allowed us to exchange views on and discuss the aesthetics of scientific data representation; by supporting the organization of the "More than Pretty Pictures" conference in 2015 that enabled us to invite researchers with valuable and stimulating insight – bringing our discussions to a new level; by hosting a workshop for the authors of this volume in December 2015; and by providing the support that made it possible for us to expand the number of color image in this volume. Without the support of AIAS – from warm tea to financial support and precious staff assistance – this book would have not be possible. For supporting the "More than Pretty Pictures" conference in 2015 specifically, we are immensely grateful to the Carlsberg Foundation (CF14-0114) and the Danish Council for Independent Research (4180-00021B).

1 Visualizing the Bacterium *Streptococcus pneumoniae*

Ditte Høyer Engholm, David Goodsell, Mogens Kilian, Ebbe Sloth Andersen, Bjørn Panyella Pedersen, and Rikke Schmidt Kjærgaard

The bacterium *Streptococcus pneumoniae*, known as the pneumococcus, is the causative agent of bacterial pneumonia, otitis media, meningitis, and septicemia. It is a major global health problem, and extensive research on the bacterium has been done. However, effective means of communicating knowledge about the bacterium to the general public is lacking. Scientists often specialize in narrow research fields, even within the biology of the pneumococcus, and they focus on communicating results to other scientists through scientific journals inaccessible to the general public.

In this chapter, we describe a visual representation of the pneumococcal biology that is meant to be accessible to a lay audience. We show how molecular data is integrated into a watercolor painting showing a 2,000,000x magnified section of the cell that will attract the attention of the viewer and stimulate curiosity and involvement. We discuss design choices and communication strategies during the process of transferring highly complex scientific data into an intentional aesthetically pleasing visual representation. The particular result shown in this chapter is a test painting. The entire project will result in a series of eight new illustrations, aimed to facilitate public awareness of the bacterium and its biology, generate hypotheses, and challenge familiar and unquestioned assumptions held by scientists.

In science, visual data representations are often the very result of the scientific process and serve mainly two purposes: reflection (internal) and publication (external). Typically, scientific figures in journals are used to communicate results to other scientists, both experts and non-experts, but rarely to a broader audience with no or little scientific background.

Initiatives such as the SciArt-program funded by the Welcome Trust 20 years ago, which had a 'significant influence on the public's engagement with science', has radically improved scientific outreach (Welcome Trust, 2006). Artists have increasingly been inspired by genetics and biology since the early 1990s, and scientists are now becoming aware of the value of art in science as well (Anker & Nelkin, 2003). However, only a minority of scientists in the natural sciences has the time and resources for data representation through artistic means in addition to the more traditional laboratory-related tasks. In addition, the major advantage of scientists being responsible for the data representations is that they have an in-depth understanding of the science that may be difficult for artists, information designers, or scientific illustrators to gain and express through the representations (Anker & Nelkin, 2003; Welcome Trust, 2006).

Our aim is to increase public awareness of the pneumococcus bacterium by making the vast molecular scientific knowledge more accessible to a lay audience in the form of watercolor paintings supported by short narratives or essays explaining the biological processes. Our reasons for doing this is that, according to the World Health Organization, the pneumococcus is the fourth most frequent microbial cause of fatal infections. It is the most common cause of bacterial pneumonia, and an estimate of more than 900,000 children under the age of 5 died from pneumonia in 2013 (World Health Organization, 2014). Pneumonia is most prevalent in the developing countries, and disease outcome depends largely on host health status and treatment initiation (Gratz, Nam Loh & Tuomanen, 2015; World Health Organization, 2011). Therefore, implementation of vaccination schemes, antibiotic strategies, as well as increased public hygiene is sufficient to significantly reduce mortality and control pneumococcal disease (World Health Organization, 2011). For this reason, it is of great importance that the general public becomes informed in order to help prevent pneumococcal pneumonia.

The comprehensive scientific work on the pneumococcus has never been artistically visualized before. We hope that the visualizations of the pneumococcus presented in this study will aid awareness by attracting attention and promoting dialogue. Furthermore, because the illustrations are as accurate as possible and based on current state-of-the-art science in the field, we believe they will be valuable in expert scientific discussions as well.

Combining our backgrounds in molecular biology, medical microbiology, and data visualization, we use scientific data to create visual representations showing 2,000,000x magnified sections of the pneumococcal cell. In this chapter, we describe the making of a test painting. It illustrates intracellular narratives shown as snapshots of small portions of the cell, showing the location, size, and shape of all the macromolecules. This is in contrast to the living cell, in which everything is in constant motion and thousands of chemical reactions happen every second. Being much smaller than the wavelength of light, the illustrated proteins and other macromolecules cannot be seen directly. We rely on methods such as X-ray crystallography, allowing us to model what is not visible to the naked eye. The main challenge of this project is to use the data available for the individual proteins as well as scientific literature to construct a visual representation that is as scientifically accurate as possible. With the annotated genome (the genetic material that codes for functional proteins and RNA) of the pneumococcus as a starting point, we use data from X-ray crystallography and protein modeling to represent the majority of the individual components of the cell. The localization in the cell and interaction with other proteins is based on scientific databases and journal articles.

Although we attempt to minimize assumptions in the paintings, some degree of artistic license will be required for a project like this to be possible (Goodsell & Johnson, 2007). For example, in our test painting, the molecules are always visualized from the same angle to enable easy recognition of specific proteins at different locations in the painting. In contrast to this, the ever-moving proteins inside a living cell would be seen from many different angles, if we were able to catch a glimpse inside of it. Moreover, we would not be able to see any details in color. Therefore, the colors chosen in the painting are not based on scientific literature. As we discuss later, schematic representations of cellular landscapes generally apply a limited color palette.

Research in the pneumococcal biology was initiated more than 100 years ago, but even though much is known, a complete understanding of the biology is not yet available (Brown, Hammerschmidt & Orihuela, 2015). Therefore, one of the most important and challenging tasks in making visual representations like the one shown in this chapter is to illustrate what is currently known in a way that will condense the scientific literature enough for it to be accessible to the lay viewer (Goodsell & Johnson, 2007). We have chosen the watercolor painting as the artistic representation because, contrary to a schematic scientific diagram, it will be familiar to the lay viewer and therefore increase the accessibility of the biology, as well as the structure and shape of the individual proteins.

We expect the style to appeal to the mainstream audience because it attracts attention, invokes curiosity, and easily conveys the scientific narrative in a simplistic style that uses as little artistic license as possible. A popular scientific short story accompanies the painting to facilitate dissemination of the narrative to the general public.

Choosing the Narratives

Our team consists of a medical microbiologist, two molecular biologists, two crystallographers, and a data visualization expert. Based on a deep literature review, we decided on eight distinct narratives for our visual representations that individually told an exciting story but collectively covered the most important aspects of the pneumococcal biology.

For each narrative, we wrote brief, state-of-the-art scientific literature reviews, not accessible for a lay audience but serving as a guide for drawing a rough sketch of the painting, and short popular scientific explanations to accommodate a lay audience. The test painting presented in this chapter shows the pneumococcal transformation process. The short popular text for the painting explaining the process of transformation is shown in Box 1.1 (the numbers in brackets refer to numbers on the rough sketch Figure 1.1A, for colors see Plate 1).

Box 1.1 Painting Narrative: Short Popular Text

Peter Parker was just a normal high school student until an irradiated spider bit him and he became Spiderman. Now, imagine what actually happened is that the spider transferred some of its DNA to Peter Parker. At the wound site, the DNA from the spider was injected into Peter's blood. The foreign DNA was incorporated into his own genes. Due to his new spider genetic material, Peter Parker developed spider abilities. In similar ways, the pneumococcus is able to take up genetic material and incorporate it into its own genes. This process is called transformation. The acquisition of new genes that encode specific traits (or abilities) is random but very important in the ongoing evolution of the pneumococcus.

For pneumococci to take up foreign DNA, a chain of events must be initiated. In the cell membrane, receptors (1–2) transmit outer signals into inner actions. A small messenger (3) functions as the outer signal, initiating the inner action: the synthesis of the first round of proteins known as the early proteins (1–6). The early proteins include more messengers (3) and they prepare the cell for the uptake of DNA. Another of the early proteins is a transcription factor (4) that mediates the synthesis of more than 80 late proteins. Three of these late proteins are responsible for the release of foreign DNA (7–9), and several proteins are required for the uptake of the released DNA (10–15).

The pneumococcus displays long, sticky, hair-like protrusions called transformation pili (11) on the surface. Each pilus is composed of thousands of small protein subunits (10). Before assembly, a preparation protein (12) trims the pilus subunits (10). Two proteins (13–14) are responsible for the assembly of the trimmed pilus subunits. The foreign DNA (17) is negatively charged and binds positive patches of the pilus (11) – much like negatively charged dust binds your positively charged computer screen.

Two proteins close to the pilus shaft (15–16) prepare the DNA (17) for entry through a pore (18) in the cell membrane, unwinding the DNA and removing one of the strands. When it gets inside the cell, DNA protective proteins (19) bind the DNA, before specific proteins (20–21) insert the foreign DNA into the pneumococcal genome.

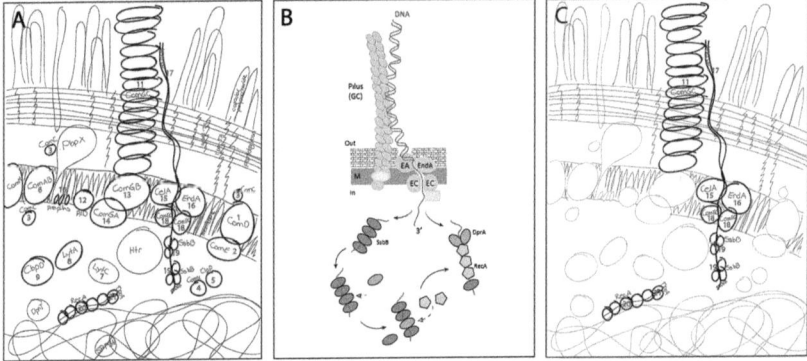

Figure 1.1 (See Color Plate 1) **A.** Rough working sketch of a section of a pneumococcal cell showing the uptake and integration of foreign DNA, known as transformation. Structures with bold outlines and red numbers are described in the short painting narrative in the box. The red line on the surface of the cell membrane marks the boundary between inside and outside. Structures above the red line are surface structures outside the pneumococcal cell, whereas the proteins and DNA shown below the red line are located inside the cell. The numbers in the sketch refer to numbers in the brackets in the text. This rough sketch serves as an overview and as a guide for drawing the final sketch for the painting (see Figure 1.4, Plate 2). **B.** This figure shows how DNA (red) binds the transformation pilus before entry proteins (labeled EA and EndA) process it. The DNA is imported through a channel in the membrane (labeled EC) and bound by protective dark green proteins and integrated in the genome by the blue and yellow proteins. The figure is reprinted with permission from the review *Streptococcus pneumoniae, le transformiste* by Johnston et al. (2014). **C.** Modified version of A, highlighting the proteins shown in both A and B. The comparison between B and C exemplifies how we used traditional journal article illustrations in composing our rough sketch. Browsing through illustrations in the literature, we quickly identified which important proteins to illustrate. Using the proteins that were illustrated most often, we broadened the literature reading to be able to compose the full narrative.

Making a List of Proteins for the Painting

The species *Streptococcus pneumoniae* encompasses many thousands of strains. Each strain represents a genetic variation of the pneumococcus. Their distinct genomes have been generated by genetic transformation and accumulation of mutations through evolution. Most of these strains express one of the more than 90 distinct capsular polysaccharide structures. As a template for our visual review, we chose the so-called TIGR4 strain because it is considered a reference organism, and it is one of the most widely used pneumococcal strains for research. Furthermore, its genome is fully sequenced, and pneumococcal experts have manually assigned the 2,240 genes to a function. Most of these genes encode proteins important for the biological function of the cell. Our first task was to decide which proteins to include in the painting.

While writing the brief scientific literature reviews, we documented the proteins in a spreadsheet. The proteins considered necessary for communicating the narrative were located in the pneumococcal genome by using a list of the annotated TIGR4 genome with locus numbers, gene symbols, and common names as well as a TIGR4 RAST spreadsheet with complete protein sequence information for all proteins (Aziz et al., 2008; Tettelin et al., 2001). The article 'Streptococcus pneumoniae, le transformiste' (Johnston et al., 2014) was one of the key review articles for the transformation narrative. We use Figure 1.1B to exemplify how we made the list of proteins for the painting. Similar to the rough sketch (Figure 1.1A), this figure shows how DNA is bound to the transformation pilus, marked in Figure 1.1B as GC. The two proteins EA and EndA then prepare it for import through the membrane channel EC. Inside the cell, the three proteins DprA, RecA, and SsbB bind the DNA and integrate it into its genome. Figure 1.1C is a modified version of Figure 1.1A, highlighting the components relevant for comparison with Figure 1.1B. The more technical aspects of identifying the proteins to be illustrated are explained in Box 1.2.

Box 1.2 Example of Protein Identification Using Databases

Each of the illustrated proteins in Figure 1.1 (Plate 1) was located in the annotated genomic list to find the genomic locus number. The cellular localization was found by using the Sybil database or the UniProt KnowledgeBase. We first identified the proteins in the annotated genome list and found their individual genomic locus number. All the illustrated proteins in Figure 1.1C except one were identified directly. The missing protein (labeled EA, an abbreviation for ComEA in Figure 1.1B) was identified by using the protein's name "ComEA" in the protein search form at the Sybil *Streptococcus pneumoniae* Comparative System Database. The search resulted in eight protein entries from different pneumococcal strains with identical protein lengths and amino acid sequences. The sequence of the amino acids in a protein is unique. Therefore, by using this amino acid sequence from one of the entries, we were able to identify the locus number of the protein in a list containing both the genomic locus number and amino acid sequences of all proteins in the pneumococcal genome. Searching the annotated genome list for the location number of ComEA obtained from the Sybil database, we identified the genomic locus number SP_0954 and its alternative name CelA labeled (15) in Figure 1.1A.

Cellular Location of the Proteins

Inside the cell, proteins are associated with specific cellular environments – some are found in the bulk interior, known as the cytoplasm, whereas others can be found on the surface. To ensure correct placement of the proteins, we used annotations of subcellular location for protein entries at the UniProt KnowledgeBase and the Sybil *Streptococcus pneumoniae* Comparative System Database. As an example, the subcellular localization of the DNA processing protein called EndA (number 16 in Figure 1.1A) is described as "cell membrane, single-pass membrane protein" by UniProt and simply "cytoplasmic membrane" by Sybil. Although both UniProt and Sybil identified EndA as a membrane protein, they provide different levels of information. In a cell, proteins can be associated with the membrane in different ways. The protein can be embedded within the membrane or only partly with a small fragment inside the membrane with a membrane anchor or with a part of it traversing the entire width of the membrane, or it can merely be close to the membrane but not physically attached to it. The Sybil annotation "cell membrane" does not distinguish between the mode of association. The annotation by UniProt, on the other hand, clearly identifies the proteins to have a fragment that traverse the membrane width. Based on its function, processing DNA before its entry into the cell, the bulk of the protein model is placed on the outer side of the membrane (number 16 in Figures 1.1 and 1.4). In some cases, the annotated localization was not used in the painting. For instance, even though a particular protein, such as the one used to build the long transformation pilus (number 11 in Figures 1.1 and 1.4) is annotated as synthesized in the cytoplasm by UniProt, we knew from the scientific literature that the bacterial life cycle required this particular protein to be localized in both cell membrane and as a pilus structure on the cellular surface.

Visualizing Proteins

To visualize proteins, structural biologists use atomic models and 3D reconstructions derived from X-ray crystallography, nuclear magnetic resonance, and electron-microscopy data. To visualize protein structures and models for the test painting, we used the molecular graphics and modeling program PyMOL. Figure 1.2 shows how a protein, in this case the pilus protein (number 11 in Figure 1.1A), can be visualized using different renderings. The default visualization mode is the line representation (Figure 1.2A). PyMOL shows the protein in the four colors green, blue, red, and white representing carbon, nitrogen, oxygen, and hydrogen atoms, respectively. For clarity, however, colors have been omitted in Figure 1.2. The advantage of showing a protein using a line representation is that the researcher can see inside the protein and how the different amino acids of which it is built are connected. Within the field of structural biology, this style is known from the painting of the protein myoglobin by the molecular artist Irvin Geis from 1961 (Protein Data Bank, 2015; Schmidt Kjærgaard, 2011). Using a maple leaf as an analogy of a protein, the line representation visualizes the midrib and the veins (Figure 1.2A lower illustration).

Depending on your interest in the protein, a ribbon diagram may be useful (Figure 1.2B). It shows the overall path of how the amino acids in the protein are organized. In the maple leaf analogy, this representation shows only the midrib and the largest veins. An elaboration of the ribbon diagram is the cartoon representation (Figure 1.2C) originally developed by biophysicist Jane Richardson (1981). A cartoon

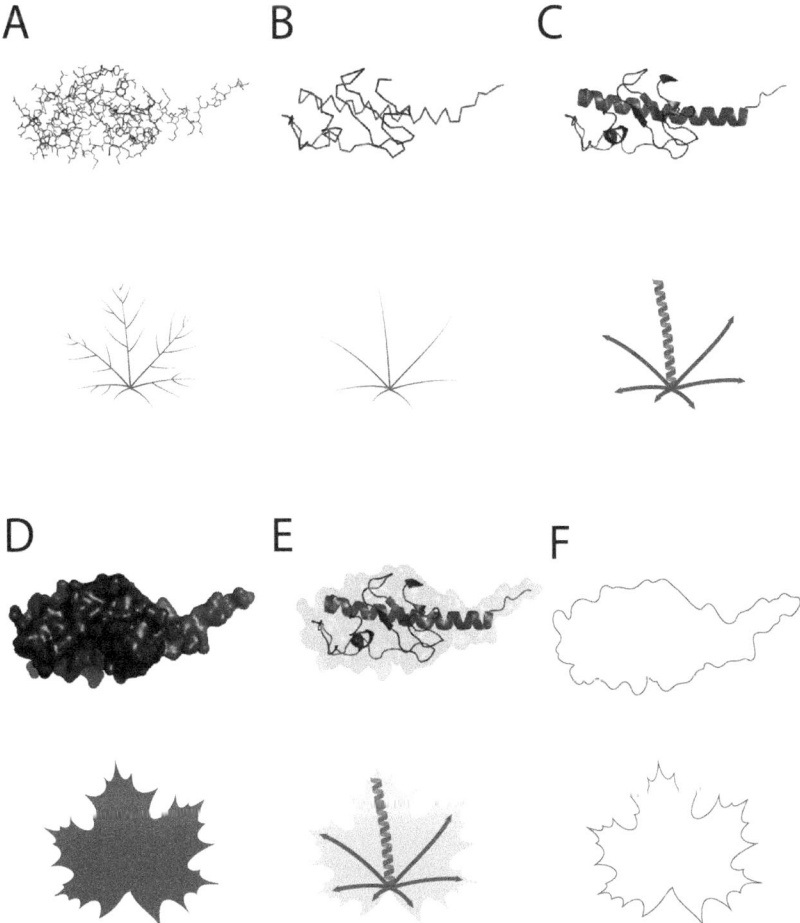

Figure 1.2 Visualization of proteins compared to a maple leaf. **A.** Line representation of the pilus protein (number 11 in Figure 1.1A) compared to the midrib and veins of a leaf. **B.** Ribbon diagram compared to the midrib and largest veins in the leaf. **C.** The cartoon representation is the most popular visualization of proteins by structural biologists. This representation illustrates specific structural features of the ribbon diagram. **D.** The space-fill representation is our best option for showing the surface of the protein. It is comparable to showing the surface of the leaf. **E.** The cartoon and space-fill representations overlapping and showing that only the space-fill, and not the cartoon, representations provide information about the outline, size, and shape of the protein in space. **F.** Our rendering of the proteins is an outline of the space-fill model as compared to showing the outline of the leaf.

representation serves to show the overall organization of the protein structure and illustrates how the protein can change shape according to function (Richardson, 2000). This representation is the preferred mode of visualization among structural biologists because it allows them to explore the structure in detail. Using this representation mode in the maple leaf results in visualizing the midrib as a spring-like string. The largest veins are shown as arrows.

For our purpose, however, a so-called space-fill model is the optimal visualization because it provides the full outline of the entire structure (Figure 1.2D). The space-fill model is an approximation of the electronic surface of the molecule. In reality, this surface may change over time due to the movement of electrons, but this representation is our best tool to illustrate the volume of the cell. It corresponds to showing the surface of the maple leaf (Figure 1.2D lower illustration). By showing the surface of the protein, it is possible for us to illustrate how proteins interact and take up space inside and outside of the cell (Goodsell, 2005). Also, a lay audience is not familiar with illustrations focusing only on the inside of an object. For instance, most people would draw the outline if asked to draw a leaf. They would know that it has "lines" inside and many will draw them, but if anyone draws only the midrib and veins it is surely a plant biologist. Furthermore, the cartoon representation (Figure 1.2C) would not have allowed us to place two molecules at the right distance to each other in the painting because we would not have been able to see where one started and another ended. Drawing two leaves as close to each other as possible requires an outline of both and not only the midrib and veins (Figure 1.2E lower illustration). Based on these considerations, we chose to illustrate the proteins with an outline of their space-fill representation (Figure 1.2F).

However, structures are not available for all proteins. In those cases, the structures can be approximated by modeling software such as SWISS-MODEL generating three-dimensional models of protein structures based on structures of proteins with similar amino acid sequences. This process is known as homology modeling (Arnold et al., 2006; Bordoli et al., 2009). The automated modeling mode provided by SWISS-MODEL generated models for almost every protein we requested. Based on the experimentally derived and homology-modeled atomic structures, we created a visual catalogue of pneumococcal proteins. PyMOL was used to display each molecule in three orthogonal views. As described above (Figure 1.2), we used the space-fill representation, since it best represents the shape and bulk of each protein. The catalogue images were created at a consistent magnification of 2,000,000x to allow direct use in our design of the pneumococcal cell painting.

Making the Final Sketch for the Painting

With a narrative, a rough sketch, and protein models prepared, we were ready to begin the watercolor painting. The space-fill images in the visual catalogue of the pneumococcal proteins were used to make the hand-drawn final pencil sketch of the narrative. The use of a smooth contour is well established in similar illustrations of cellular environments (Goodsell, 1991, 2005, 2009; Goodsell & Johnson, 2007; Voet & Voet, 2009). We explored the effect of a bumpier contour similar to the visual representations in PyMOL (e.g. Figure 1.3). The intention was to emphasize the atomic structure of the proteins and give the impression that the viewer would see more detail if the magnification was increased, i.e. that you would be able to see the individual amino acids if the magnification was large enough. At the same time, the contours are simple enough that they do not distract from the understanding of the overall arrangement and interaction of molecules. The bumpy contours enhance the flickering sensation of the uncertainty of visualizing something that is otherwise invisible to us. The contours also imply that a closer look at the proteins would reveal more about their atomic nature. In the final painting, black outlines indicate how the macromolecules interact. Molecules that act individually have a black outline, and assemblies of several molecules share an outline.

We drew all protein complexes on separate paper to fine-tune the rendering of each of them (Figure 1.3) before we transferred all sketches to the painting (Figure 1.4, Plate 2). The transformation pilus spiral structure was based on a type on a similar pilus structure solved by Craig et al. (2006). As depicted in Figure 1.3A the pilus protein is assembled into a helical structure. Another example of a protein complex is the so-called ABC transporter. This particular ABC transporter is composed of three components, known as subunits in molecular biology, with two of them being

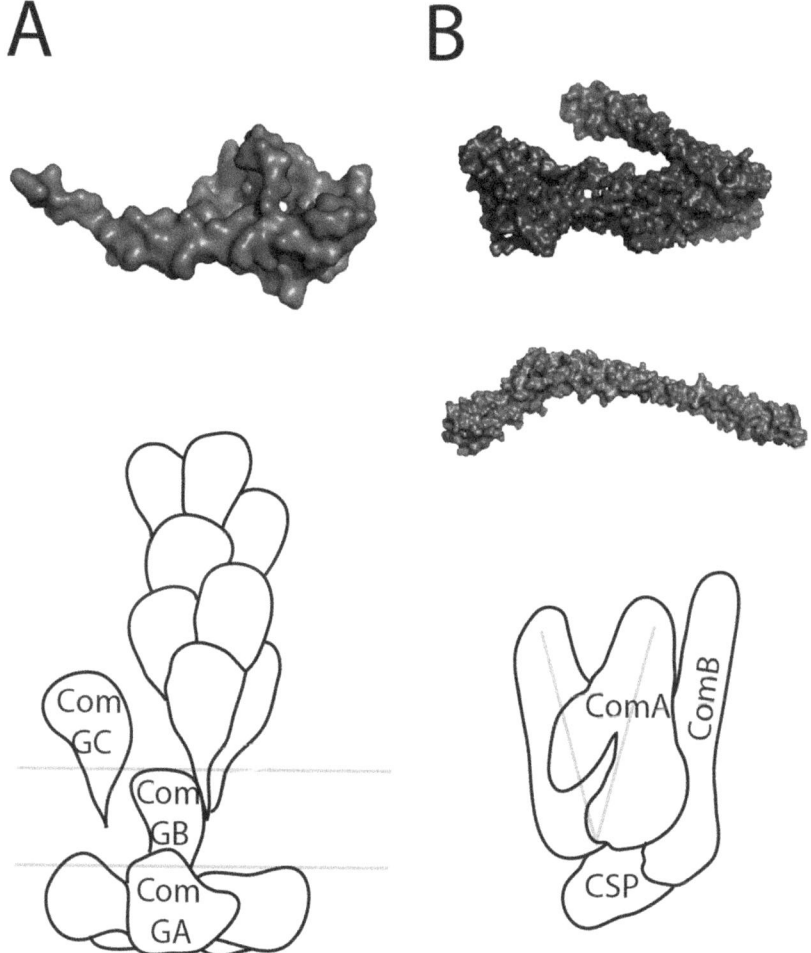

Figure 1.3 Preparation for the final drawing. **A.** The upper part is a space-fill representation of the pilus protein (labeled number 11 in Figure 1.1A). The lower illustration is our rendering of the transformation pilus consisting of several pilus proteins in a helical conformation based on the article by Craig et al. (2006). **B.** Visualization of the complex that transports messenger across the membrane (labeled number 6 in Figure 1.1A). The lowest part of the figure shows the illustration of the complex based partly on the article by Shintre et al. (2013).

identical (upper part of Figure 1.3B). Based on a recent crystal structure of an ABC transporter (Shintre et al., 2013), the two identical subunits are shown as a V-shaped complex (the gray V-shaped indication in the lower figure) to indicate the flipping conformation change that transports the messenger to the outside of the cell. The other subunit (middle part of Figure 1.3B) is known to be closely associated with the transporting part of the complex, and both are required for the transport of the messenger (Hui & Morrison, 1991; Hui, Zhou & Morrison, 1995). Therefore, this stalk-like protein is shown in close contact with the transporting part on its right-handed side. The assembled protein complex is shown in the lower part of Figure 1.3B.

With all components ready for the final sketch for the painting, we decided on an overall composition. To facilitate easy interpretation of the painting, we placed the transformation pilus in the top-left corner to lead the eye of the viewer into the painting. We chose a diagonal placement of the membrane to underline the dynamic environment inside the cell, where proteins are constantly moving (The J. Paul Getty Trust). The final painting sketch, containing the foreground of the painting only, was transferred to watercolor paper. The black bar in the lower right corner corresponds to 5 centimeters in the painting and 25 nanometers in the cell (Figure 1.4A, Plate 2).

The Painting Process

With the final pencil sketch transferred to the watercolor paper, we decided to try to make the painting read as a coherent whole, as it might look as if we were actually able to look inside the pneumococcal cell through a microscope. Therefore, we chose a water tone uniform colored palette. In previous works, where the aim has been to highlight functional compartments within the cell, the color palette has been more diverse (Goodsell, 2009, 2012).

Recent illustrations used for journal articles show an overweight of a pale brown color palette to illustrate the interior of the cell (e.g. Medina & Garcia-Sastre, 2011; Mitchell, 2003; Short, Murdoch & Ryan, 2014). In the field of molecular biology, where plants are valuable model organisms, a green cytoplasm may indicate that the illustrated cell is in fact a plant cell. As an example, a Google search for "plant cell illustration" results in mainly green illustrations of plant cells. Despite the trend in choosing a pale brown color for the cytoplasm of non-plant cells, we explored cold hues of blue and green in the painting because the green and blue color tones invite the viewer to spend more time exploring the details of the picture because of the immediate monotony of the colors.

The foreground was painted first in clear colors as exemplified by structures in the cell wall (Figure 1.4B, focus 1, see Plate 2 for colors). For key proteins, a lighter tint of green was used in the cytoplasm to make them the most prominent features in the final painting, accentuating the narrative (the pale green in focus 2). To create a sense of depth, proteins in the middle ground were colored in a darker shade of the same colors (the two darker shades of green in focus 3), and proteins or structures in the background were even darker (illustrated by the very dark blue in focus 4). Similar to the space not occupied by proteins, the outlines of complexes and individual proteins were colored black (focus 5).

All proteins except the blue membrane-bound are in shades of green. The key proteins for the transformation narrative are highlighted in light tints of green in the cytoplasm. However, in the nucleoid region, they are not as prominent due to a smaller

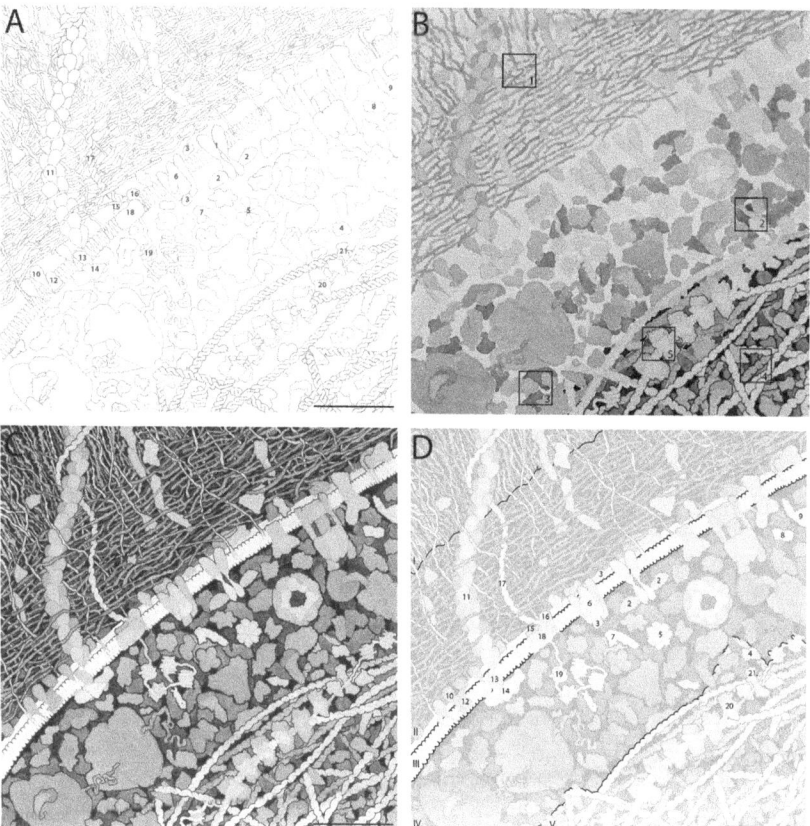

Figure 1.4 (See Color Plate 2) Painting pneumococcal transformation. **A.** Final painting sketch showing the foreground of the painting. The proteins and DNA described in the painting narrative (Box 1.1) are labeled accordingly in red. The black scale bar to the lower right represents 5 centimeters in the painting and corresponds to 25 nanometers in the cell. **B.** This illustration shows the process of the painting. First, the foreground is painted (focus 1). The proteins labeled with red in A are highlighted by using a very light tint of green and blue (focus 2). Darker proteins are added in the middle ground (focus 3), and even darker proteins or DNA, as shown here, are added in the background (4). Black contours are used to outline complexes where two or more proteins or DNA interact (5). **C.** Final pneumococcal transformation painting. Photo by Michael Grøn, Billedmageren – Grøn og Grønborg Fotografi. **D.** Visual explanation for the final painting. The black numbers are identical to the numbering in A indicating the localization of key proteins. From the top left toward the lower right, the painting shows the capsule (labeled I) and the cell wall (II) on the surface of the cell. The cell membrane (III) encloses the interior of the cell, which consists of the cytoplasm (IV) and the nucleoid region (V). Photo by Michael Grøn, Billedmageren – Grøn og Grønborg Fotografi.

contrast to the light blue DNA. Neither the pilus proteins in the cell membrane nor the cell wall are easily recognizable either. The intention with the chosen color palette was to make the painting a coherent whole. However, by using a very light blue tint for the cell membrane and the DNA, these structures become the most prominent features of the crowded cytoplasm of the painting. These features are more noticeable than intended, and the painting comes across as having three compartments in addition to the membrane: the nucleoid region, cytoplasm, and extracellular region, where the capsule and cell wall appear coherent despite being blue and green, respectively. The overall impression of the painting is still coherent, although there is a subtle impression of the cellular compartments (Figure 1.4C). In the middle- and background, the gradually darker and less saturated colors work well to give a sense of depth.

Pneumococcal Transformation 2,000,000x

The final painting shows other aspects of the biology of the pneumococcus than merely the transformation process. It shows the structure of the capsule (section I in Figure 1.4, Plate 2) made of long carbohydrate molecules and the sieve-like structure of the cell wall (section II). Inside the cell membrane (section III), the cytoplasm (section IV) is packed with proteins, and in the nucleoid region (section V), the DNA excludes most proteins. So far, the test painting has received positive responses from both scientists within the field of molecular biology as well as from a lay audience. Molecular biologists acknowledge the benefits of the painting.

Viewers without a background in molecular biology have been inspired to ask questions in order to understand what they see. The painting stimulates curiosity and inquisitiveness in a way that a text without illustrations would not have been able to (Ware, 2008). Experts of pneumococci are initially skeptical toward sketches of the organism that they gather molecular information about. However, once they realize that the sketches are based on solid scientific and molecular data, they are surprised by the density of proteins in the bacterial cell and by the relative sizes of its component structures. Therefore, the painting may provide even experts with new insight about their area of research.

Observations for Future Work

The painting presented in this chapter contributed valuable experiences before we initiated the process of painting a series of eight selected pneumococcal narratives. The data collection process and protein modeling will be used for future paintings. The process described is standardized and will serve as a template for generating similar cellular visualizations. The interplay between manual and automated work is optimal for making a representation as accurate as possible in a reasonable amount of time. The output of the data collection process is the protein models each viewed from four different angles in our protein catalogue. We made a hard-copy print of the catalogue to use directly when drawing the sketches for the painting. When printed at 62%, the proteins are at a consistent 2,000,000x magnification. This magnification allows us to consistently and easily paint and recognize the individual protein.

For future paintings, we will create compositions that are more easily read, i.e. we will place proteins in a strictly left to right manner according to reading direction. Furthermore, proteins that act consecutively in a biological pathway and are mentioned together in the narrative are placed close to each other, if possible. To make sure that the membranes illustrated in the paintings are realistically curved, we will start the making of the sketches by drawing a 2,000,000x magnified pneumococcal cell. We will trace the projection of a microscopy image on a wall to serve as a guide for the membrane curvature.

We will use the bumpy contours explored in the painting presented here for future paintings as well. Using the protein catalogue, we will trace the outline of the proteins and transfer these directly to the sketch. Similar to the rendering in the test painting, we will indicate subunits in protein complexes using a slightly darker tone.

The water-inspired color palette used in the painting invites the viewer to spend more time exploring it, but to facilitate recognition of the cellular compartments, we will employ a diverging color scheme, although more subtle than previously used (Goodsell, 2009). The future color palette will be inspired by a brown-to-green diverging color scheme from ColorBrewer (Harrower & Brewer, 2003). Among the colorblind-safe palettes, we found the three shown in Figure 1.5 (Plate 3) most aesthetically pleasing.

Inspired by the water and earth tones, we chose a brown-to-green color scheme. Also, the difference between the outer cold green tones nicely complements the warmer brown tones on the inside of the cell. Using a diverging color palette of green and brown, we have chosen a cold purple color for the key proteins. Based on a color analysis of both chroma and lightness of the green and brown tones, we will select a saturation of the purple color that is equally different from the other colors. In this way, we will make sure that the key proteins will be the most prominent features in the paintings.

Our test painting and the future eight pneumococcal paintings lie on the border of art and science, and they encompass both expressive and inexpressive components. The inexpressive components include the nature of the proteins that we illustrate because we cannot choose the shape, size, and localization of the proteins. However, the test painting is mostly expressive because we have made objective choices about the representation of the protein models, as well as the color and its symbolism. Furthermore, by illustrating what is not visible to us, this test painting is our representation of a nano-scale world in the form of the pneumococcus.

Although the paintings may be classified as art based solely on its expressive components, our main aim is to increase public awareness of the bacterium with our series

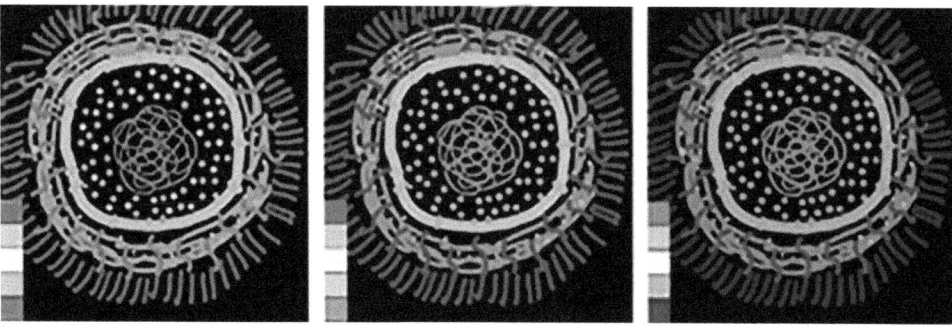

Figure 1.5 (See Color Plate 3) Choosing colors for the painting series. Based on the diverging colorblind-safe color palettes from ColorBrewer.org, we selected the three schemes most pleasing to us as indicated by the color bars in the lower left corner of each illustration. The three illustrations shown here served as a basis for which to decide on a general color palette for the eight future paintings.

of paintings illustrating the pneumococcus. As mentioned, simple precautions can help prevent and treat the majority of infections. To the public, we intend to present the eight final paintings each accompanied by a version with faded colors and numbers (like Figure 1.4D, Plate 2) and the short popular narratives since we hope that the combination of the paintings and accompanying narratives will attract the attention of a lay audience, which has shown no prior dedication to biological sciences. In this way, we also hope to increase the interest in science through artistic means.

References

Anker, S. and D. Nelkin. 2003. "Deciphering DNA: The Art and Science of a Supermolecule." In *The Molecular Gaze: Art in the Genetic Age*, edited by P.R. Reilly, pp. 1–7. New York: Cold Spring Harbor Laboratory Press.

Arnold, K., L. Bordoli, J. Kopp, and T. Schwede. 2006. "The SWISS-MODEL workspace: A web-based environment for protein structure homology modelling." *Bioinformatics* 22(2): pp. 195–201. doi:10.1093/bioinformatics/bti770.

Aziz, R. K., D. Bartels, A. A. Best, M. DeJongh, T. Disz, R. A. Edwards, K. Formsma, S. Gerdes, E. M. Glass, M. Kubal, F. Meyer, G. J. Olsen, R. Olson, A. L. Osterman, R. A. Overbeek, L. K. McNeil, D. Paarmann, T. Paczian, B. Parrello, G. D. Pusch, C. Reich, R. Stevens, O. Vassieva, V. Vonstein, A. Wilke, and O. Zagnitko. 2008. "The RAST Server: rapid annotations using subsystems technology." *BMC Genomics* 9: p. 75. doi:10.1186/1471-2164-9-75.

Bordoli, L., F. Kiefer, K. Arnold, P. Benkert, J. Battey, and T. Schwede. 2009. "Protein structure homology modeling using SWISS-MODEL workspace." *Nat Protocols* 4(1): pp. 1–13. doi:10.1038/nprot.2008.197.

Brown, J., S. Hammerschmidt, and C. Orihuela. 2015. Preface to *Streptococcus pneumoniae— Molecular Mechanisms of Host-Pathogen Interactions*, edited by J. Brown, S. Hammerschmidt & C.J. Orihuela, pp. xiii–xvii. Cambridge: Academic Press.

Craig, L., N. Volkmann, A. S. Arvai, M. E. Pique, M. Yeager, E. H. Egelman, and J. A. Tainer. 2006. "Type IV pilus structure by cryo-electron microscopy and crystallography: Implications for pilus assembly and functions." *Molecular Cell* 23(5): pp. 651–662.

Goodsell, D. S. 1991. "Inside a living cell." *Trends in Biochemical Sciences* 16(6): pp. 203–206. doi: 10.1016/0968-0004(91)90083-8.

Goodsell, D. S. 2005. "Visual methods from atoms to cells." *Structure* 13(3): pp. 347–354. doi:10.1016/j.str.2005.01.012.

Goodsell, D. S. 2009. *The Machinery of Life*, Second Edition. Göttingen: Copernicus Books.

Goodsell, D. S. 2012. "Illustrating the machinery of life: Viruses." *Biochemistry and Molecular Biology Education* 40(5): pp. 291–296.

Goodsell, D. S. and G. T. Johnson. 2007. "Filling in the gaps: Artistic license in education and outreach." *PLoS Biology* 5(12): p. e308. doi:10.1371/journal.pbio.0050308.

Gratz, N., L. Nam Loh, and E. Tuomanen. 2015. "Pneumococcal Invasion: Development of Bacteremia and Meningitis." In *Streptococcus Pneumoniae—Molecular Mechanisms of Host-Pathogen Interactions*, pp. 433–451, Chapter 23. Cambridge: Academic Press.

Harrower, M. and C. A. Brewer. 2003. "ColorBrewer.org: An online tool for selecting colour schemes for maps." *The Cartographic Journal* 40(1): pp. 27–37.

Hui, F. M. and D. A. Morrison. 1991. "Genetic transformation in *Streptococcus pneumoniae*: Nucleotide sequence analysis shows comA, a gene required for competence induction, to be a member of the bacterial ATP-dependent transport protein family." *Journal of Bacteriology* 173(1): pp. 372–381.

Hui, F. M., L. Zhou, and D. A. Morrison. 1995. "Competence for genetic transformation in *Streptococcus pneumoniae*: Organization of a regulatory locus with homology to two lactococcin A secretion genes." *Gene* 153(1): pp. 25–31.

Johnston, C., N. Campo, M. J. Bergé, P. Polard, and J. P. Claverys. 2014. "*Streptococcus pneumoniae, le transformiste.*" *Trends in Microbiology* 22(3): pp. 113–119. doi:10.1016/j.tim.2014.01.002.

Medina, R. A. and A. Garcia-Sastre. 2011. "Influenza A viruses: New research developments." *Nature Reviews Microbiology* 9(8): pp. 590–603.

Mitchell, T. J. 2003. "The pathogenesis of streptococcal infections: From tooth decay to meningitis." *Nature Reviews Microbiology* 1(3): pp. 219–230.

Protein Data Bank. 2015. "Geis Digital Archive." Cited 2016 March 7. http://pdb101.rcsb.org/geis-archive/irving-geis.

Richardson, J. S. 1981. "The anatomy and taxonomy of protein structure." *Advances in Protein Chemistry* 34: pp. 167–339.

Richardson, J. S. 2000. "Early ribbon drawings of proteins." *Nature Structural Biology* 7(8): pp. 624–625.

Schmidt Kjærgaard, R. 2011. "Things to see and do: How scientific images work." In *Successful Science Communication—Telling It Like It Is*, edited by D. J. Bennett and R. C. Jennings, pp. 332–354. Cambridge: Cambridge University Press.

Shintre, C. A., A. C. Pike, J. I. Li, Q. Kim, A. J. Barr, S. Goubin, L. Shrestha, J. Yang, G. Berridge, J. Ross, P. J. Stansfeld, M. S. Sansom, A. M. Edwards, C. Bountra, B. D. Marsden, F. von Delft, A. N. Bullock, O. Gileadi, N. A. Burgess-Brown, and E. Carpenter. 2013. "Structures of ABCB10, a human ATP-binding cassette transporter in apo- and nucleotide-bound states." *Proceedings of National Academy of Sciences of USA* 110(24): pp. 9710–9715. doi:10.1073/pnas.1217042110.

Short, F. L., S. L. Murdoch, and R. P. Ryan. 2014. "Polybacterial human disease: the ills of social networking." *Trends in Microbiology* 22(9): pp. 508–516.

Tettelin, H., K. E. Nelson, I. T. Paulsen, J. A. Eisen, T. D. Read, S. Peterson, J. Heidelberg, R. T. DeBoy, D. H. Haft, R. J. Dodson, A. S. Durkin, M. Gwinn, J. F. Kolonay, W. C. Nelson, J. D. Peterson, L. A. Umayam, O. White, S. L. Salzberg, M. R. Lewis, D. Radune, E. Holtzapple, H. Khouri, A. M. Wolf, T. R. Utterback, C. L. Hansen, L. A. McDonald, T. V. Feldblyum, S. Angiuoli, T. Dickinson, E. K. Hickey, I. E. Holt, B. J. Loftus, F. Yang, H. O. Smith, J. C. Venter, B. A. Dougherty, D. A. Morrison, S. K. Hollingshead, and C. M. Fraser. 2001. "Complete genome sequence of a virulent isolate of *Streptococcus pneumoniae.*" *Science* 293(5529): pp. 498–506. doi:0.1126/science.1061217.

The J. Paul Getty Trust. "The J. Paul Getty Trust—Elements of Art." Cited 2015 October 15. www.getty.edu/education/teachers/building_lessons/formal_analysis.html.

UniProt. "DNA Ligase LigA—entry Q97QT2." Cited 2015 July 14. www.uniprot.org/uniprot/Q97QT2.

UniProt. "UniProt KnowledgeBase." Cited 2015 August 17. www.uniprot.org.

Voet, J. G. and D. Voet. 2009. "Communication through illustration: The work of David Goodsell." *Biochemistry and Molecular Biology Education* 37(4): p. 203. doi:10.1002/bmb.20303.

Ware, C. 2008. *Visual Thinking for Design.* Burlington: Morgan Kaufmann Publishers.

Wellcome Trust. 2006. "Insight and Exchange: An Evaluation of the Wellcome Trust's Sciart programme." Cited 2015 September 3. www.wellcome.ac.uk/About-us/Publications/Reports/Public-engagement/Sciart-evaluation-report/index.htm.

World Health Organization. 2011. "Pneumococcal Disease." Cited 2015 July 9. www.who.int/immunization/topics/pneumococcal_disease/en/.

World Health Organization. 2014. "Pneumonia." Cited 2015 July 9. www.who.int/mediacentre/factsheets/fs331/en/.

2 Diagrammatic and Iconic Imagery in Science

Frederik Stjernfelt

The role of diagrams in science has been evident (at least) since Euclid, and diagram use spans both theoretical and practical sciences (for example, geometry and cartography), just like it covers natural, social, and humanity disciplines. This chapter elaborates on the difference between diagrams and icons, and discusses the diagrammatic and iconic elements at work in the case of the watercolor painting of the bacterium *Streptococcus pneumoniae* presented in Chapter 1.

The prototypical diagram is a 2D graphical representation of some state-of-affairs, providing a skeletal selection of important relations within that state-of-affairs. Icons are signs that, in some respects, are similar to the objects they refer to – paintings, photos, images, but also diagrams, algebras, etc. Correlatively, iconicity refers to the general quality of referring by means of similarity. Among icons, diagrams form a specific type where the emphasis is placed on the clear, often explicit relations between the parts of the icon – which makes it possible to reason about the character of those relations. There is a gradual continuum between images and diagrams, and while the former may play important scientific roles (e.g. astronomical photography, brain imaging), the latter are especially crucial for representing scientific results because of their potentiality for expressing general results.

What Is a Diagram?

Typically, the diagram not only selects some aspects of its object only, but it also stylizes and generalizes those aspects selected. To the American logician and philosopher Charles Sanders Peirce (1839–1914), an important feature in diagram reading was the abstraction process, which he called "prescission" (Peirce, 1992, 1998a, 1998b; Stjernfelt, 2007). This implies that certain features of the diagram token on the page, table, or screen are abstracted away by the reader in his/her mind in order to be able to focus on the essential content represented in that token. For example, a triangle in a geometry textbook has a series of properties that must be bracketed in order to be properly understood: The triangle token, e.g., has a color that is not the case for the general triangle; it has a particular size; its angles have particular values – all this must be bracketed in order for the type of triangle to be envisioned. This also goes for the fact that the lines of triangle tokens have a certain breadth – while lines in Euclidean geometry are defined by having zero breadth – hence, breadth must also be subject to prescission. Thus, there is a whole series of cognitive operations, which the reader of a diagram must undertake – typically without reflecting much upon it – in order to address the general object depicted by the diagram. In a certain sense, the converse operation is that of "filling-in" – when you have a general or abstract representation, for instance a diagram – you can re-concretize that representation by adding, in the imagination or otherwise, additional particular properties not represented in the diagram, thus creating a more particular operation (Ingarden, 1973; Stjernfelt, 2015a).

The prescission procedure then paves the way for the most important property of the diagram according to Peirce – once the reader accesses the diagram type, he or she is able to draw conclusions *by experimenting* upon the diagram, most often in imagination, sometimes by the adding of further lines or marks to it, sometimes by physically changing the relations between its parts (Stjernfelt, 2007, 2014, Chapter 10; Pietarinen, 2006). Such experimentation, of course, is dependent upon certain rules, which differ from one diagram to the next and which are informed by the ontology of the object depicted. By adding lines and manipulating the triangle after the basic set of rules characterizing Euclidean geometry, it is possible to prove theorems, such as the famous theorem that the angle sum of any triangle equals the sum of two right angles (180°). In other diagrams, experiments may include the finding of a route on a map, the solving of an equation or – as in the example of Chapter 1 – the transformation process allowing for pneumococcus bacteria to partially modify their DNA sequence.

Thus, in the Peircean view, the pragmatic use of diagrams is not that of communication only – if by communication is meant the forwarding of some intellectual content already established. It is also to present a vehicle for further thought experimenting – in certain cases, the reader may even be able to use the diagram to conduct thought experiments that were never foreseen by the manufacturer or the sender of the diagram. This cognitive aspect of the diagram is based on the fact that it may contain relations between its parts, which have not yet been noticed but which further use may make explicit. In that sense, diagrams also furnish public, common devices for thought.

Like texts, diagrams are often shaped according to the creator's imagination of the capabilities of its supposed readers. Thus, in scientific texts, the reader addressed is most often a scientific colleague, supposedly with a scientific erudition similar to that of the author and thus able to understand, process, and experiment with the diagram at a level similar to that of the author. Such scientific diagrams are very often central to the paper in which they occur. It has been claimed that such diagrams are mere supplementary illustrations of what is more properly expressed in the accompanying text, but in many cases, this is not true. Very often, the diagram synthesizes a large mass of empirical data into one figure; in other cases, it combines those data with central theoretical assumptions – and in many cases, the diagram simply expresses, in a synthesizing form, the central empirical finding or theoretical claim advanced by the paper in question.

Diagrams used in popular science do not differ in any principal way from the diagrams used by and among scientists. Typically, they may make more explicit aspects that are tacitly understood by scientists; they may simplify scientific diagrams by leaving out parts, levels, or aspects; or they may combine diagrammaticity with more immediately iconic representations or more explicit symbolic instructions in ordinary language – or all of these at the same time. Simultaneously, "popularization" is not a simple process. It varies considerably depending on the intended recipient – from fellow scientists in neighboring fields or more remote faculties to the informed readers of high-level science journals like *Scientific American* or more popular journals like the English *BlueSci*, the American *Wired* or *New Scientist*, over less-informed laymen seeking information about certain scientific results, e.g. for epidemiological reasons and to school kids or kindergarten children. Hence, "laymen" are not all the same but come in many categories with very different levels of general and scientific information and diagram reading skills.

The Pneumococcus Examples

Figure 1.1A (Plate 1) is presented by the authors of Chapter 1 as a rough sketch, but it actually functions as a very good example of a diagram. Apart from numbers and protein names, it consists of line drawings only, using the colors gray, black, and red to serve distinct purposes. Gray and black form contours of partial objects on each side and across cell surface, back- and foreground, respectively, while the red line serves the purpose of indicating the decisive inside-outside boundary of the pneumococcus cell. Some of the distinct objects in the foreground and a few in the background indicated by closed lines are labeled with an abbreviated name, and the foreground objects, in addition, are numbered from 1 to 21 – numbers referring to types of object indicated in a legend list. The sketch-like character of this diagram – cf. the "roughness" indicated by its creators – is given by the sloppy line layout. No two closed lines indicating objects have the same shape or size, and some of them even have repeated lines demarcating their boundary by approximation.

This may be compared to the textbook equivalent in Figure 1.1B (Plate 1), where related objects are stylized to have the same shape and size and where their different types are given with different, codified coloration. If we compare the two, it may seem like the sketch does not work as a diagram, but thanks to the labeling and numbering, however, this lack of type indication in the drawing is compensated, the types being clearly indicated by the non-iconic means of numbers and labels. Despite not being very colorful or in other senses "appealing", it appears to be a very concise and easy-to-understand representation of the process depicted. In order to serve its purpose to be accessible to a lay audience, the diagrammatic and rather scientific elements of the preliminary sketch (Figure 1.1A, Plate 1) were downplayed in the final test painting (Figure 1.4C, Plate 2) in favor of more iconic elements in order to create a more inviting version.

In this transformation process from Figure 1.1A to Figure 1.4C in Chapter 1, several means are employed, each of them "re-iconizing" or "filling in" the diagrammatic sketch in Figure 1.1 (Plate 1) in distinct ways in the sense that they re-introduce features that promote iconic meaning. One is to introduce a varied color palette of different green, bluish-green, greenish-blue, and blue shades applied in watercolor. Watercolor is an iconic medium in the sense that the hues rarely are exactly the same, which provide a watercolor painting with a more vibrant snapshot-like style than, for instance, standard, saturated focal colors. Watercolor generally offers a more flickering impression than oil paint (in art, it is suitable for mimicking the sense of shifting light in outdoor scenery). The idea to use watercolor as a medium for representing a lively and ever-changing bacterial environment makes sense, but the iconic style challenges modern natural sciences' conventional preference for diagrammatic images, which typically favors use of easily identifiable focal colors, like focal black, red, blue, green, yellow, etc. – each such color is used to identify one distinct object or one distinct type of object. The (water) color scheme and style of Figure 1.4C (Plate 2) thus provides the image with iconic dimensions that allow the viewer to be intrigued by the visuals without revealing what the image is meant to represent – or even that it is meant to represent something.

An example of a different and strictly diagrammatic style and color scheme would be the famous London underground map devised by Harry Beck in 1931 and used as a matrix for numerous subway maps around the world ever since. Here, focal colors are used to identify subway lines so that each line has its own color, easily distinguishable from the others, so that it is cognitively quick to "prescind" (Peirce's term; Peirce 1998a, 1.353) one line from the otherwise indistinct spaghetti of intertwined lines on the map. Another example is the standard chemical building blocks where atoms are represented by circular balls characterized by size and a focal color for each element (Hydrogen white, Oxygen red, Nitrogen blue, Carbon black, etc.). In the London underground case, *individual* objects – subway lines – are distinguished by focal colors; in the molecule case, *types* of objects – elements – are distinguished by the same means.

The application of color in Figure 1.4C (Plate 2) differs from such standard diagrams: The objects depicted – the submicroscopic cell parts – cannot be said to have any colors in themselves (like subway lines or atoms, neither of which has colors identifying them), so any color choice can be applied to them. By using a series of different intermediary shades between blue and green, Figure 1.4C implicitly communicates to the viewer that he or she should read the figure as an image and not as a diagram.

Another non-diagrammatical feature introduced in Figure 1.4C (Plate 2) is the giving up – or all but giving up – of the foreground-background distinction of Figure 1.1A (Plate 1). Now, all partial objects are "on the same level", making it much more difficult to know which part of the picture the reader is supposed to focus on. This is anti-diagrammatical in the sense that it dispenses from the normal economy of diagrams where relevant diagram parts are clearly placed in focus and accentuated vis-à-vis the background (see, for instance, the diagrams analyzed in Chapter 3). Conventional diagrammatic features, however, are given to two other versions of the same image, Figure 1.4A giving the gray outlines of Figure 1.4C with all the colors subtracted but the 1–21 numbering of Figure 1.1 reintroduced. The two of them then recombine into Figure 1.4D, which reintroduce the colors but in a faded version so that the numbers are still visible. In any case, this step makes Figure 1.4 (Plate 2) a much more "full" picture with an almost horror vacui quality so that every plane segment over a given granularity is taken up by some compound. This is no doubt a much more iconic representation of the actual, complicated, compact, hard-to-overview conditions of actual cell biochemistry than the diagrammatic Figure 1.1A (Plate 1).

A further anti-diagrammatic feature, finally, is that of introducing molecule shape to the different chemical compounds. This is done by means of the introduction of stylized versions of the bunch-of-grapes-like representations of molecules as ball clusters with smoothed-out surfaces. Contour line drawings of such representations (indicated by Figure 1.3) now represent single compounds in Figure 1.4. The clusters do not convey any diagrammatic information, but this partial reintroduction of iconicity provides the viewer with the simple, yet significant, impression that there are in fact many molecule types floating around. It gives the impression of Figure 1.4C as being a small sample of an indefinite space of chemical complexity stretching far beyond the boundary of the actual image.

As a quasi-abstract watercolor, Figure 1.4C (Plate 2) has aesthetic qualities such as shape variety, color hue, harmony, and spatial density – all of them lacking in its "rough" predecessor Figure 1.1 (Plate 1), which on the other hand had the diagrammatic feature of cognitive clarity. It is still retaining the diagrammatical structure of Figure 1.1, but the aesthetic qualities added give a diminishing of perspicuity, of immediate understandability, and of cognitive clarity as a result.

Cognition and Beauty in Images

It is an open question whether the appealing aesthetic qualities achieved in Figure 1.4C (Plate 2) "outweigh" the loss of clarity in Figure 1.1A (Plate 1). If Figure 1.1A scares away science-frightened potential readers otherwise attracted by Figure 1.4C, then Figure 1.4C might constitute a gateway to understanding, which Figure 1.1A does not. If, on the other hand, the less easy parsing of Figure 1.4C into its relevant diagram pathway parts prevents "lay audience" from actually understanding the process, little is gained. This problem is solved by displaying to the audience *both* images in parallel (as intended by the authors of Chapter 1), since it is generally the case that when you understand several different representations of the same thing or state-of-affairs, you understand the object better than from one representation only. Double or triple representation of the same thing allows the observer to "triangulate" the properties of the objects. Also, as the "general public" is constituted by rather different layers of class, gender, age, education, etc., those not caught by one representation might be caught by another.

But as to the priorities when constructing the single image, the examples of Figures 1.1A (Plate 1) and 1.4C (Plate 2) make it clear that sometimes beauty and cognitive clarity may have functions, which calls for careful consideration of purpose and audience in each case. The matter, however, is made more complicated by the fact that beauty is no simple quality. The beauty of watercolor shadings and complicated shapes may be aesthetically pleasing to our senses – just like the stylization of focal colors might be – but there also exists such a thing as *the aesthetics of cognition* or *the aesthetics of knowledge* (Chapters 4 and 6 also elaborate on this). Take, for example, the Beck map of the London underground. Not only has this (very stylized) diagram become a hallmark of the city of London and has been copied in its principles by urban transportation systems all over the world, but it also enjoys a considerable aesthetic success. Numerous caricatures and even artworks take the Beck map as their origin and substitute station names for other taxonomies of words. The acquisition and construction of knowledge is connected to certain pleasures, far from all of them utilitarian or otherwise immediately means-end oriented. Rather, the achievement of certain insights, results, or aims has an aesthetic quality of its own (not unlike many other practices from cooking to trekking), which is probably part of why scientific diagrams may be seen as possessing aesthetic qualities and possibilities. Thus, the fact that diagrams, like matrices, tables, equations, architects' designs, construction recipes, etc., may possess a beauty is widely recognized by many different artists, such as Nikolaus Gansterer, Matthew Ritchie, and Robert Talbot (see Whittle, 2014).

But if diagrams may possess, in themselves, a series of aesthetic possibilities in their own right, it would be too simplistic to frame the problem in a duality of *clarity vs. aesthetics*, as might immediately be supposed from the tension between Figures 1.1A (Plate 1) and 1.4C (Plate 2) of Chapter 1. Then, it would rather be an issue of *different kinds* of beauties, some inherent in or supporting diagrammatic representations and others working against it. This can be seen from comparing the initiative of Chapter 1 with one of the classics of twentieth-century diagram making, the Vienna positivist Otto Neurath's 1939 *Modern Man in the Making*. Like the authors of Chapter 1, he also wanted to reach out to the general public – to ordinary modern human beings in democracies, cf. his title. His argument was that in order for the general public easily to understand facts and problems of demography, economics, sociology, welfare, and so on and thus make informed decisions in democratic elections, they needed to have *diagrammatic representations* of such facts and problems, represented in diagram maps, column diagrams, pie charts, and pictograms using simple shapes and simple focal colors. In turn, this gave rise to the diagrammatic language known as Isotype.

In a way, the authors of Chapter 1 have the same purpose as Neurath in the sense that the aim is to visually represent complex matters. They use different means, however. Neurath used simple shapes and focal colors to directly appeal to the common man and ease his understanding. They use the clash between subtle (abstract) water-color and text narratives to evoke curiosity. In modern society, it is a long time ago now that aesthetic preferences were stably shaped by academies, intellectuals, leading museums, and artists. Rather, aesthetic sensibility is sure to differ widely across age, class, gender, ethnicity, geography, and much more – and one and the same person may be able to shift swiftly between very different aesthetic attitudes facing different objects. Hence, in order to reach the many-headed ghost called "the general public" and teach it about the dangers of pneumococcus, the use of a *variety* of aesthetic means is important.

References

Ingarden, R. 1973. *The Literary Work of Art*. Evanston: Northwestern University Press (1931).

Peirce, C. 1992. *The Essential Peirce*: *Selected Philosophical Writings, Volume 1 (1867–1893)*, edited by Nathan Houser and Christian Kloesel. Bloomington and Indianapolis: Indiana University Press.

Peirce, C. S. 1998a. *Collected Papers, I–VIII*, edited by C. Hartshorne, P. Weiss, and A. Burks, pp. 1931–1958. London: Thoemmes Press.

Peirce, C. 1998b. *The Essential Peirce: Selected Philosophical Writings, Volume 2 (1893–1913)*, edited by the Peirce Edition Project, 1998. Bloomington and Indianapolis: Indiana University Press.

Pietarinen, A. V. 2006. *Signs of Logic: Peircean Themes on the Philosophy of Language, Games, and Communication*. Dordrecht: Springer.

Stjernfelt, F. 2007. *Diagrammatology. An Investigation on the Borderlines of Phenomenology, Ontology, and Semiotics*. Dordrecht: Springer Verlag.

Stjernfelt, F. 2014. *Natural Propositions: The Actuality of Peirce's Doctrine of Dicisigns*. Boston: Docent Press

Stjernfelt, F. 2015a. "Green War Banners in Central Copenhagen: A Recent Political Struggle Over Interpretation—And Some Implications for Art Interpretation as Such." In *Investigations into the Phenomenology and the Ontology of the Work of Art. What are Artworks and How Do We Experience Them?*, edited by P. Bundgaard and F. Stjernfelt, pp. 209–224. Cham: Springer.

Stjernfelt, F. In press. "Schematic Aspects of an Aesthetics of Diagrams." In *The Temptation of the Diagram*, edited by M. Ritchie. Los Angeles: Getty Research Institute.

Whittle, M. 2014. "Romantic Objectivism: Diagrammatic Thought in Contemporary Art." PhD diss., Kyoto City University of Arts. Available at: www.michael-whittle.com/uploads/1/1/1/3/11134083/romantic_objectivism.pdf. Accessed 28 April 2017.

3 Scientific Data Visualization

Aesthetic for Diagrammatic Clarity

Martin Krzywinski

'The great tragedy of science – the slaying of a beautiful hypothesis by an ugly fact', wrote the English biologist Thomas Henry Huxley, also known as Darwin's Bulldog, in a statement that is as much about how science works as about the irrepressible optimism required to practice it. Equally great is the tragedy of obfuscating beautiful facts with ugly and florid visuals in impenetrable figures. It is not only a question of style – systemic lack of clarity, precision, and conciseness in science communication impedes understanding and progress.

Well-designed figures can illustrate complex concepts and patterns that may be difficult to express concisely in words. Figures that are clear, concise, and attractive are effective – they form a strong connection with the reader and communicate with immediacy. These qualities can be achieved by combining data-encoding methods, to quantify, with principles of graphic design, to organize and reveal. Best practices in both aspects of visual communication are underpinned by conclusions from studies in visual perception and awareness and respect how we perceive, interpret, and organize visual information (Cleveland & McGill, 1985; Heer & Bostock, 2010).

Scientific figures are often judged by how well they function as a diagram. As accounted for in Chapter 2, diagrams should present data neutrally, responsibly include uncertainty and sources of potential bias, achieve contrast between patterns that are meaningful and those that are spurious, encapsulate and present derived knowledge, and suggest new hypotheses.

Equally important, though more elusive, is a figure's aesthetic, visual engagement, and emotional impact. What is the relationship between this visual form and underlying function, and how can any such connection be used to formulate best practices? Is the form of a scientific figure entirely dictated by its function, or can form be decoupled and independently altered? This chapter addresses these points specifically with regard to diagrams.

Visual Grammar

We all use words to communicate information – our ability to do so is quite sophisticated. We have large vocabularies, understand a variety of verbal and written styles, and effortlessly parse errors in real time. But when we need to present complex information visually, we may find ourselves at a "loss for words", graphically speaking.

Do images and graphics possess the same qualities as the spoken or written word? Can they be concise and articulate? Are there rules and guidelines for visual vocabulary and grammar? How can we focus the viewer's attention to emphasize a point?

Can we modulate the tone and volume of visual communication? These and other questions are broadly addressed through design, which is the conscious application of visual and organizational principles to communication. Their practice can be thought of as a kind of visual rhetoric whose aim is to clearly transmit ideas and experimental outcomes without bias while maintaining intellectual integrity and transparency.

All of us have already been schooled in "written design" (grammar) and most of us have had some experience with "verbal design" (public speaking), but relatively few have had training in "visual design" (information design and visualization). However, before we learn *how* to visually communicate, we must figure out *what* we wish to say.

Constraints and Clarity

The scientific process works because all its output is empirically constrained. As such, the reporting of scientific knowledge can be dry and unemotional. R. M. Pirsig writes in his novel from 1974 that the purpose of science is 'not to inspire emotionally, but to bring order out of chaos and make the unknown known', underscoring the fact that clear communication is paramount (Pirsig, 1974). This is because "order out of chaos" rarely arises from a single observation or theory but from many small incremental steps, and, as such, it is important to clearly communicate the minutiae, not only the grand discoveries, to improve the chances of scientists to successfully connect ideas.

This deep requirement for clarity and specificity means that both written and visual style must be 'straightforward, unadorned, unemotional, economical and carefully proportioned' (Pirsig, 1974) because 'rich, ornate prose is hard to digest, generally unwholesome, and sometimes nauseating' (Strunk Jr., 1920). It is 'not an esthetically free and natural style. It is esthetically restrained. Everything is under control' and the quality of communication 'is measured in terms of the skill with which this control is maintained' (Pirsig, 1974). When information is hastily arranged or tinted with arbitrary personal taste, its impact and fidelity can easily be diluted.

The primary payload of scientific communication is its information content. The form of the communication must therefore always be subordinate to it. Form must not only respect content but also elevate it, clarify it, and untangle its complexity. Anytime it overwhelms content, we are likely to find signs of bad design choices and possibly lack of respect for information. The footprint of design should therefore be subtle, despite the fact that the product may hastily be perceived as possessing 'surface ugliness' (Pirsig, 1974) because it lacks the loud and tawdry design tropes used to please the eye and disengage critical thinking made familiar by marketing and advertising.

That being said, unfamiliar forms condemned as "overwhelming content" may actually be the products of genuine and inventive explorers who create new ways of respecting content and connecting us to it. Such hasty perceptions should be avoided and judgment deferred until the motivation and intention of the creator are well understood. For example, although the form of the pneumococcal transformation in the watercolor painting (Figure 1.4 in Chapter 1, Plate 2) is radically different from what would normally appear in the primary scientific literature (Figure 1.3), it should never be regarded as having less utility. In addition to stimulating a wider audience to participate in the conversation about the importance of pneumococcus, it demonstrates a key aspect of the biochemistry that is missing from traditional depictions, namely, that cells are packed with biologically active molecules to a much greater extent than would normally be inferred from traditional figures. This dense packing may be surprising and is a convenient springboard to discussion about critical concepts such as entropy, entropic trapping, kinetics, and thermodynamics.

Design Is Choreography for the Page

When visualizing a complex process, concept, or data set, it is irresponsible to dump the information on the page and leave the reader to sort it out. We understand enough of how visual information is organized and processed to realize that often our readers' visual system can sabotage our attempts at communication.

The human visual system, from retina to cortex, is extraordinarily complex. But even basic knowledge of its working can be leveraged to avoid surprises – "you saw what in my figure?" For example, knowing how people perceive color allows us to create data-to-color mappings in which relative changes in data can be visually perceived with accuracy. This mitigates the optical illusion of the luminance effect, in which the perception of a color is influenced by nearby colors (Wong, 2010a). We also know that some things grab our visual attention first (and don't let go; Wong, 2010b, 2011a; Yantis, 2005), that the distribution patterns in collections of similar shapes elude us (Krzywinski & Wong, 2013), and that we cannot accurately judge relationships between areas (Heer & Bostock, 2010). These disparate observations about our visual system are brought together into a phenomenological model known as the Gestalt principles (Wong, 2010c, 2010d), which can guide how we organize elements on a page to achieve flow and salience that is compatible with what we're trying to communicate.

It is more productive to think about visual communication from the top down rather than the bottom up. The main motivating factor informing design decisions should be the core message (what are we trying to say?) and aspects of its quality such as accuracy, clarity, conciseness, and consistency. With these in mind, we can approach design as a balance of focus, emphasis, detail, and salience. Only once these points are addressed can we begin to reflect on the task of communicating from the bottom up and make practical selections about data encoding, color, symbols, typeface, arrows, line weight, and alignment.

Where possible, design should depict familiar physical processes as intuitively as possible. It would be irrational to graphically represent the addition of a reagent to a test tube from the bottom with the tube pointing down (Figure 3.1A). Yet this is exactly what we're expected to accept by the quantitative timeline depiction (Araldi, Ferrari & Levine, 2015). Once the tube and reagent are drawn right-side-up, a more appropriate encoding of the timeline readily presents itself (Figure 3.1B). This simple flip is part of the subtlety of the aesthetic.

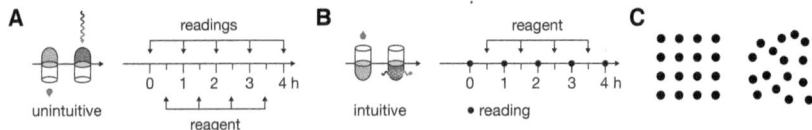

Figure 3.1 **A.** An unintuitive graphical and corresponding quantitative depiction of adding a reagent and carrying out a measurement (Araldi, Ferrari & Levine, 2015). **B.** An intuitive presentation of the experiment in Figure 3.1A. **C.** Controlling spatial variation greatly assists subitizing, the act of quickly identifying the number of elements in a group without counting. Determining that there are 16 dots is effectively instant when they are arranged on a grid.

We should always aim to lower the cognitive load of parsing a figure. To achieve this, we can make use of our ability to subitize – a process in which we can determine the number of shapes in a group without explicitly counting them (Wender & Rothkegel, 2000). Our ability to perform this kind of pre-attentive counting is limited to about 3–5 elements, and there is diverse evidence in how it relates to explicit counting (Vetter, Butterworth & Bahrami, 2008). Spatial variation (Figure 3.1C) disrupts this process.

Just as this kind of arbitrary spatial variability should be limited, aspects of color, shape, and texture should also be controlled – when unnecessarily varied, they compete for visual attention. At all times, visual salience must be informed by relevance – the figure should not be a visual distraction theme park (Wong, 2010b, 2011a). For example, excessive visual flourish makes it more difficult to present a narrative with ordered elements because we cannot reliably predict the path of the reader's eye – everything looks important, and it is not clear where to look (Figure 3.2A, Plate 4). By adopting minimal and unencumbered elements of style (Krzywinski, 2013), formatting aspects such as space, color, and shape gain specificity (Figure 3.2B, Plate 4). It is difficult to use spot color to emphasize similarity in a figure replete with color. For example, it would be difficult to make element B in Figure 3.2A stand out because there is already too much competition for our attention. If other elements could be made more subtle, such as in Figure 3.2B, it is enough to spot color with red. Salience can be achieved with even subtle markings. The white dot inside the rectangles of elements C and D in Figure 3.2B are very quickly seen – the bounding rectangle acts as negative space to focus our attention on the dot – and allow us to indicate a relationship between C and D.

An even more important contributor to clarity in design is the requirement that elements vary in proportion to the variation in the data or concepts they represent. This relatively strict requirement comes from the fact that quantitative aspects of shapes, such as position, size, shape, and color (hue, richness, luminosity), are all data encodings. When these aspects are decoupled from data and allowed to change based on arbitrary design choices, it becomes difficult to assess which differences are due to data and which are due to formatting. For example, each of the elements A, B, C, D, and E in Figure 3.2A (Plate 4) have not only horizontal but, because of their shapes, also vertical extent. What should we infer from the fact that D is taller than A? Elements A

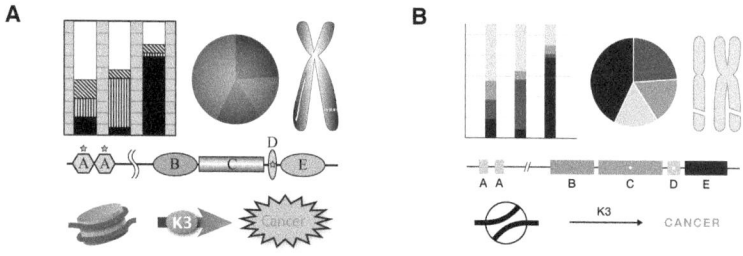

Figure 3.2 (See Color Plate 4) **A.** Florid visuals are distracting and make adding emphasis difficult. **B.** Conservative formatting of shapes from Figure 3.2A more effectively present spatial and semantic relationships between elements and make adding emphasis easier.

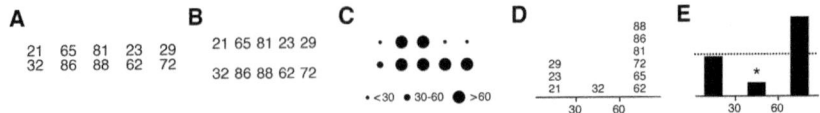

Figure 3.3 **A.–E.** To reduce the number of unrelated and unproductive interpretations, detail in data should be controlled to match the core message, showing aggregate statistics to emphasize relevant properties (e.g. count within value ranges).

and D share color and a star but have different shapes. How is the star different from the color? Should we infer similarity between B, D, and E because they are ovals?

Limiting variation is equally important in quantitative figures. Practically, because the core message in quantitative information is often difficult to pin down – there may be several equally important points to raise or the data set may not yet be fully understood – there is a tendency for authors to rush straight into the choice of data encoding and hope that the relevant data will be fortuitously arranged and informative patterns will somehow present themselves. Hoping that the data will speak for themselves is a dangerous gamble, even in small data sets (Figure 3.3A–E). When raw data are shown, there remain many possible interpretations, which may be motivated by formatting and layout. Grouped numbers into pairs (Figure 3.3A) motivates within-pair comparisons leading to the observation that the second number in the pair is always larger and that the first number begins with the same digit that the second one ends. Grouping numbers into rows (Figure 3.3B) makes it easier to observe that the first row is composed of odd numbers and the second row of even numbers, something that was harder to spot in the pair grouping. Without guidance from the figure designer, the reader cannot know whether these observations are relevant. By discretizing the numbers into one of three ranges, the importance of specific values is played down (Figure 3.3C). Showing the histogram of the counts of numbers (Figure 3.3D) emphasizes that counts within ranges are important and makes it obvious that the odd/even distinction is not important – the reader has been subtly steered away from unproductive thoughts about the data. Finally, by removing the values altogether (Figure 3.3E), the core message is distilled further: There are significantly fewer mid-sized values than expected.

Every element in Figure 3.3E is necessary to convey the message (removing the dashed line deprives the reader of information needed to fully formulate the conclusion). Clarity is achieved because detail in the data has been encapsulated to expose only the relevant differences, much like the detail in fanciful shapes in Figure 3.2A has been suppressed in Figure 3.2B, which avoids visual clutter. Note also that the use of the * to mean "statistically significant" is a visual convention in data statistics that requires no explanation in the figure itself (the legend should describe the statistical test, sample size, P value, and effect size). Such conventions are powerful – the figure becomes self-contained.

It is not enough to merely maximize the data-to-ink ratio (Tufte, 1992), which is the proportion of ink (or pixels) that encode the data, but have in mind which patterns can be discerned and what conclusions offered by the figure are actionable. Thus, actionable data-to-ink ratio is also important – it is not enough to show what we have; we need to have some sense of what can be done with it.

As important as these concepts are, without a means to practically and consistently implement them in figures, they remain mere hopes, and effective communication becomes a happy accident. Below, I use examples from the literature to show how design choices affect these concepts and to demonstrate that minor changes in formatting can greatly improve comprehension in the "concept figure" and the "data figure".

The Concept Figure

The concept figure is one of the most challenging to design. Concepts can be based on progression, change, transformation, relationships, and differences – all of which could be spatial, temporal, qualitative, or both. Matching visual properties, such as shapes, color, and position, of elements in the figure to unambiguously and specifically match these relationships requires attention.

To think about visual design of concepts, we must fall back to the fundamental concepts of top-down design discussed in the context of Figures 3.1–3.3. Is the core of the concept clear? Are key points found along the figure's central vertical and horizontal axes? Is the eye encouraged to flow along an intuitive path of explanation? Is there sufficient encapsulation to shelter the reader from extraneous details? Is relevant similarity between objects made clear? Have things that are dissimilar been accidentally grouped by shape, color, or proximity? If there is a temporal flow or spatial compartmentalization, is the progression clear? If the same phenomenon is shown at different scales, is the organization of the magnified vignettes compatible with the rest of the figure?

Part of Figure 3.4A (Plate 5) was used in Chapter 1 as the starting point to generate a watercolor representation of the molecular biology of *Streptococcus pneumoniae*. The final painting beautifully depicts the complexity, density, and diversity of the molecular agents both in the process and background. Design decisions were balanced by knowledge about the system and personal aesthetic, and, critically, the latter was never allowed to override the former. The image encourages and rewards exploration and, when accompanied by an interpretive legend, has pedagogical value.

If we look back to the original concept figure from literature (Johnston et al., 2014), we find that it similarly contains both knowledge and personal aesthetic. However, the sophistication of the aesthetic lags behind the information content in the figure – ad hoc and inconsistent design interfere with clear presentation. The cluttered design introduces ambiguities, which make understanding the concepts difficult: Why is the host chromosome arranged in three loops? What does the brick pattern in the membrane represent? What is the square under the right EC molecule? Why are some arrows dashed? What is the significance of different shapes of proteins (pentagon, oval, circle)? These ambiguities compound to make the entire process difficult to grasp without resorting to the legend.

The multiplicity of colors and shapes creates a mismatch between salience (what stands out to the eye) and relevance (what is important). Background entities and navigational components should complement and not compete with the core message. For example, yellow is a very salient color and is used to highlight the EndA protein, which degrades the double-stranded DNA, a critical step. Unfortunately, RecA is similarly colored. This not only takes focus away from EndA, but it also motivates similarity grouping and suggests proteins are somehow related, but there is no evidence in the paper that they are.

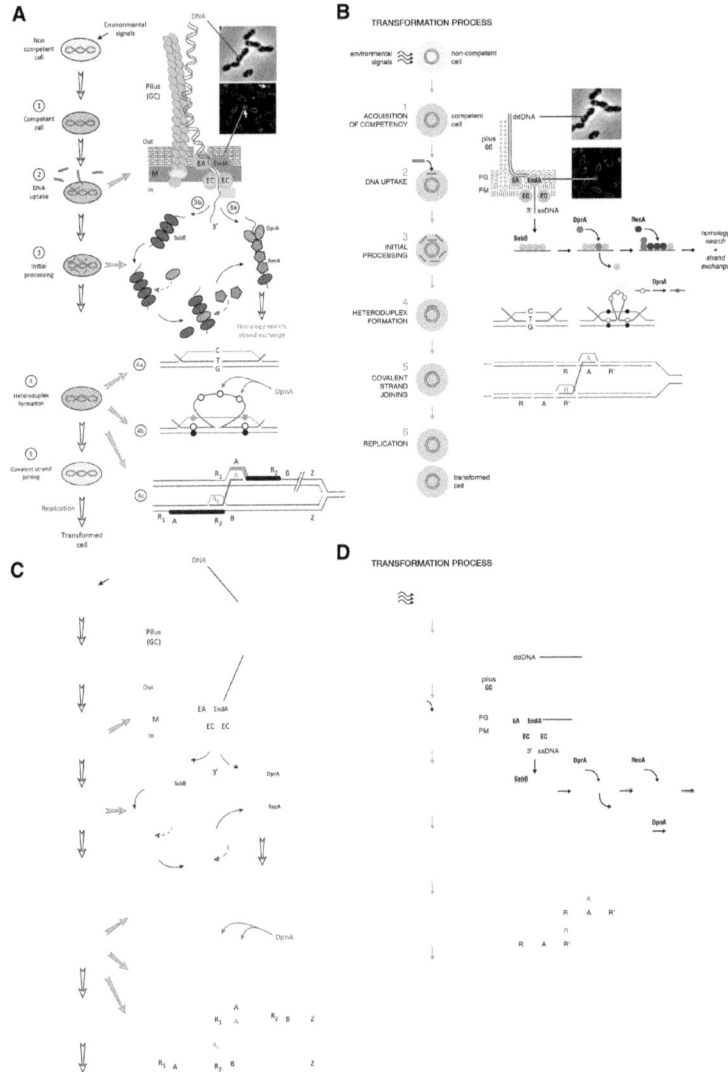

Figure 3.4 (See Color Plate 5) Consistent alignment and judicious use of color and shape variation helps convey complex concepts and processes. **A.** Figures 1 and 2 from Johnston et al. (2014). **B.** Their redesign making use of alignment, simple shapes, and increased focus on the transforming dsDNA. **C.** Arrows, callouts, and entity labels from A. **D.** Arrows, callouts, and entity labels from B.

The presence of green, used to color the host DNA, the host cell, and proteins reduces the salience of red. More importantly, the green/red combination is not colorblind-safe – these colors will appear similar to readers with deuteranopia, and all focus on DNA will be lost (Wong, 2011b).

A redesign of Figure 3.4A is presented in Figure 3.4B (Plate 5). Variation in shape and color is minimized. The green host color has been replaced by grey – removing color from background elements increases the salience of the red used for the DNA strands and makes it easier to perceive changes along the pathway. Blue has been retained speculatively for DpnA because the protein appears in steps 3 and 4 – any such emphasis should be justified in the legend. All the strands have uniform thickness (0.75 pt) and are drawn as simply as possible – it is unnecessary to mimic the double helix motif, or any kind of bending.

Topologically complex processes like strand joining should be shown as simply as possible. The choice of equal lengths, consistent spacing, and alignment helps identify similarity between elements, such as sequence similarity that makes step 5 possible. In the original figure, it is not clear where the start and end of the labeled sequence regions (A, B, R1, R2) are. Moreover, it's not clear whether R1 and R2, identified as repeat sequence, are actually identical or different instances of the same repeat family. Only A, R1, and R2 need to be identified in the process – both B and Z are superfluous. The redesign presents the two double strands as identically as possible, with simplified labels for the repeat sequence, R and R′, instead of the cumbersome subscripts. By keeping all labeled segments the same length and placing the red ssDNA in the center of the figure, the focus is kept on the bridging process.

One of the ways in which the cohesion of a figure can be assessed is to independently picture its functional and navigational components. For example, if we draw only the arrows and labels of Figure 3.4A, we find a jumble (Figure 3.4C). The environmental signal comes in right to left, which contradicts the more intuitive left-to-right direction for a beginning step. The spacing between the arrows between steps 1, 2, and 3 is inconsistent, as is the alignment between the green horizontal arrows that point to the molecular mechanisms of steps 2 and 3 (see Plate 5). Why is the first arrow angled, and why is the second green arrow shorter? Had elements been placed with more foresight, all the arrows could be made horizontal or vertical, of the same size, and consistently spaced. The arrows in the cycle of transformation in the bottom right have various amount of curvature, and their orientation does not help in establishing smooth flow. In particular, the arrow above the two RecA molecules is very disruptive.

Labels are both inconsistent and poorly aligned. For example, what is "Out" and "In"? The words suggest motion, but really they really mean "extracellular" and "intracellular" and are redundant if the membrane is indicated clearly. Labels of protein names vary in size, which breaks grouping by similarity, making it more difficult to perceive the protein name labels as objects of the same category. Curiously, the "In" and "Out" labels are larger than the protein names in the initial processing loop, which are much more relevant. The first instance of the blue protein that we are likely to see (bottom middle), because of the large amount of negative space around it, which functions to emphasize the shape, is unlabeled. The DprA label is found at the edge of the figure, quite far from where it is most needed. Note that the SsbB is labeled in its first salient instance, unlike all the other proteins.

When the arrows and labels in the redesigned figure are shown in isolation (Figure 3.4D), we see a consistently spaced backbone of navigational elements that act as a grid and assist in imposing order on information. Protein labels are bolded for emphasis and separate from other labels, and PG and PM labels have been added to indicate peptidoglycan and plasma membrane components of the cell wall.

Figure 3.4 (Plate 5) is a good example of how simplification and clarification of complex visual presentation affects its aesthetic.

The Data Figure

The data figure is different from the concept figure in that it is based on quantitative information that can be numerically transformed, can be accompanied by measures of uncertainty, and whose proportions can be mathematically compared. The data figure, in addition to the kinds of relationships and hierarchy of information found in the concept figure, may also need to capture statistical trends and show clearly uncertainty in the observation or departures from mathematical models. Ideally, a data figure should present data in a way that makes any relevant patterns clear and, just as importantly, not obscure such patterns. Depending on the figure, both concept and data components may exist. For example, a figure that explains the serial steps in an algorithm may include the kind of flow and steps in Figure 3.4, along with graphs of the output of numerical simulation.

No matter how complex the data or concept is, the corresponding figure must give access to this complexity. The reader should have the sense that, given enough time (a few minutes), they will gain familiarity with the way information is presented and have confidence that they have not missed any important trends. It is up to the figure to draw attention to any trends that are unusual or potentially subtle. For example, differences in lengths of elements can be more accurately assessed if the elements are aligned (Heer & Bostock, 2010). This access to complexity is even more important in a data figure than a concept figure because the latter tends to be read serially, much like text, whereas the former can have its elements parsed in various order. In a data figure, the reader seeks to find spatial patterns between elements, which are determined by data and not strictly design.

The choice of data encoding should not be made based on the data type. For example, if we have a data set of pair-wise relationships, we should not think that this should necessarily be shown as a network diagram. In fact, network diagrams, for example, are notoriously difficult to assess because of their unpredictability and inaccessible complexity (Krzywinski et al. 2011). Such data could be shown as an adjacency matrix or even a list (Gehlenborg & Wong, 2012). The choice of encoding should be motivated by the questions that are being asked and the types of trends that we expect to parse from the figure.

Even very simple data sets can present challenges. Although we are extremely competent in navigating the world visually, we cannot count on our visual system to assess proportions and find patterns. For example, Figure 3.5A (Plate 6) compares two Venn diagrams that show the fraction of cells in brain regions active in different actions (Iso: isometric force production, Obs: visual observation, Sac: saccadic eye movements). Each Venn diagram encodes 7 numbers, so we are asked to compare 14 numbers, all the while keeping the categories in mind as well as their totals.

A subtle property of the figures is that the percentages in neither diagram add to 100%. The difference is classified as "none" and accounts for 22% in the left diagram and 61% (nearly three times as much!) in the right – a fact that is easily missed because the Venn diagram areas are not truly proportional. We might be led to believe that the Venn diagram circles consistently show proportion because some circles are scaled in the right direction (e.g. Sac on the left is larger than Sac on the right). But this is quickly contradicted by the fact that this scaling is inconsistent (22% on the left seems to cover a smaller area than the 19% on the right,

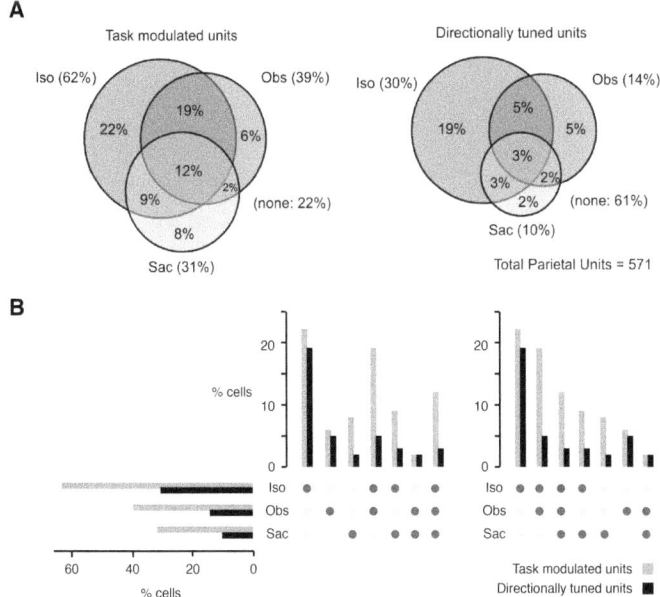

Figure 3.5 (See Color Plate 6) Data encoding should be chosen to clearly demonstrate proportion. The use of areas to represent quantity should be avoided. **A.** Adapted from
Figure 2 in Ferrari-Toniolo et al. (2015). Venn diagrams show proportion of cells
in inferior parietal brain regions are significantly modulated (left) and directionally
tuned (right) in three different tasks: isometric force production (Iso), visual observation (Obs), and saccadic eye movements (Sac). **B.** UpSet encoding of the data in
Figure 3.5A.

while 8% on the left is larger than 2% on the right). It might appear that circle
size was based on the absolute amounts, but the original figure legend description
('Percentages of cells modulated and directionally tuned') does not suggest this to
be the case.

Even though they are much simpler than network diagrams, Venn diagrams are very
difficult to compare (one could argue that network diagrams are actually impossible
to compare). Given that the purpose of the figure is to quantitatively communicate
the differences between the fractions of activated cells in different actions and units,
the Venn encoding is inadequate. In fact, it is likely that the Venn form was chosen
entirely because it is the most suitable for the data structure, without consideration of
the fact that any questions that arise are very poorly addressed by it.

The reader is left to assess differences between the data sets in the figure by explicitly
comparing the numbers in each intersection. This could be facilitated by aligning the
numbers to minimize eye travel and removing the redundant "%". Labeling lacks
consistency – the format of "Iso (62%)" is not compatible with "(none: 22%)". Again,
design choice is in direct conflict with the meaning of the information. Why is the
"none" label closer to the Venn circles than the "Obs" label, given that it is not a
category represented in the diagram (it corresponds to the fraction of cells that were

not activated in the task)? The authors attempt to deal with this by enclosing "none" in parentheses – a bad decision made to deal with a problem due to an initial bad decision. Now, the labels are inconsistent, while "none" is still close to the diagram. When faced with a situation in which compromise must be reached (how do I fit this label in this circle?) one must be on the lookout for a choice made upstream that has forced the compromise (could the circles that contain text be made to be rounded rectangles?).

One of the fundamental problems with the Venn encoding is that it does not allow alternative views of the data. The categories (Iso, Obs, Sac) and their intersections cannot be arbitrarily reordered to communicate trends in how the values increase, decrease, or accumulate. For example, it's impossible to quickly assess which of the intersection represents the middle-sized fraction in the left (Iso + Sac) and right (Iso + Sac and Iso + Obs + Sac) Venn diagrams.

Whenever comparison of multiple quantities is important, a scatter or bar plot should be considered. The bar plot more emphatically communicates values – in a bar plot, the amount of ink is proportional to the value, whereas in a scatter plot, this is encoded by the lack of ink (distance between point and axis). By encoding the information using the UpSet encoding – like Figure 3.5B (Lex et al., 2014), we can decouple the values from the categories. In this encoding, the values for each intersection are shown as a bar plot, and the intersection is identified by a matrix of symbols below. This approach allows us to reorder intersection values – we can sort by either unit type (task modulated or directionally tuned), which allows us to immediately spot trends that were previously obscured. For example, if we sort by descending values of task modulated units (left bar plot in Figure 3.5B), we see from the matrix below the bar plot that the "Iso" action is associated with the four largest values. The trends between task modulated and directionally tuned units also become apparent. In the former, the fraction of cells drops much more slowly across categories than in the latter – where the difference between the largest value (Iso, 19%) and next largest (Obs, 5%) is very large.

Differences of Differences

Many data sets are too large to show all the information that is collected in an experiment. In biology, to explore biological variation, we typically need a large number of samples. The differences between these samples may be minor and, if the data is presented in its entirety, may be lost among the similarities. For example, a healthy and a diseased cell are more similar than they are different, but it is the differences that are of interest.

For data encodings that are not quantitative, such as the sequence alignments in Figure 3.6A (Plate 7), emphasizing differences requires that we suppress the display of areas of the figure where the data do not change and make the consensus (or reference) data the central focus (Figure 3.6B). By doing this, we are saving the reader from having to find all the sequence positions across species that vary with respect to the human sequence, a process fraught with error. Given that the comparison is made to the human sequence, it should be placed first. The order of other species should be based on phylogeny or, as in Figure 3.6B, a measure of the difference of their sequence compared to human. For example, we might place the species that have perfect conservation first (cow, whale, bushbaby, and Tasmanian devil) and others in order of the total number of positions in which their sequence varies, possibly limited

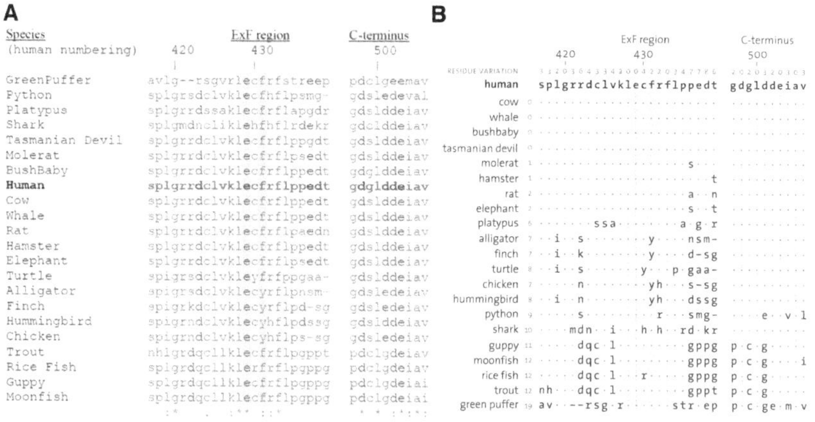

Figure 3.6 (See Color Plate 7) Focus on differences helps identify patterns in data where small variation exists within a consensus. **A.** Adapted from Figure 8 in Li et al. (2015). The figure shows the evolutionary conservation of sequence across different species. **B.** Redesign of Figure 3.6A. Emphasis on differences is achieved by showing the human sequence first and in other species only those residues that are not conserved. The extent of differences is quantified by showing the number of different residues at a given position – in the original figure, this important quantity was encoded using symbols and relegated to the bottom of the figure. The order of species is based on the number of variants.

to only those changes in which fundamental properties of the residue change (e.g. hydrophilicity).

By focusing on the differences of the sequences with respect to humans, we can quickly group species based on common (or different) differences. For example, the guppy, moonfish, rice fish, and trout at the bottom of the figure clearly show similar differences in the alignment (dqc·l, gpp, p·c·g), something that would have been very difficult to detect as quickly in the original.

Just like in the example of the Venn diagram in Figure 3.5A, where we wanted to bring related labels close to the diagram (e.g. Iso task) and distance others (e.g. "none"), here, too, the distance between labels and data benefits from adjustment. When the species names are left-aligned, their distance to their sequence is unnecessarily increased, especially in cases of short labels like "cow". The size of this effect depends on the length of the longest species label (Tasmanian devil). The figure design is thus brittle – a single label has influence over all other labels. This issue is entirely avoided by right aligning. The only time that left alignment would be helpful is if it was important that parts of the labels were vertically aligned for easy comparison, such as would be the case for alphanumeric sample names.

It is also not important for the species labels to be capitalized. In general, capitalization should be entirely avoided unless required by convention, such as gene names or sequence structures (ExF, NOS1AP). Note the inconsistent use of spacing: GreenPuffer vs Rice Fish. The Gestalt principle (Wong, 2010c, 2010d) of grouping – either by similarity or distance – is very useful and, when disregarded, can cause confusion and excess eye movement. In this figure, the labels for the species column, "Species", and sequence position row, "(human numbering)", should be moved so that both labels are closest to the part of the table they are labeling. As it is in Figure 3.6A, the column numbering header appears immediately below the species header, which makes it look like the header is one of the species. The unfortunate choice of adding parentheses to the row header also makes the text look like it is modifying the "Species" header (e.g. that human numbering applies to the species column). These problems can be mitigated by removing both of these headers – they are actually not necessary. The row labels are obviously species, and the column numbers are obviously sequence positions. The fact that the positions are relative to the human sequence does not need to be part of the figure because it is not part of the central message (it belongs in the legend).

The use of color in Figure 3.6A (Plate 7) classifies the amino acid residues by their chemical property (hydrophobic, hydrophilic, neutral, etc.), and different schemes exist for specific elements (Procter et al., 2010), though many of the conventions are not colorblind-safe nor normalize luminance. Here, color interferes with finding regions where sequence is different because hue is a much stronger discriminator and has more of a grouping effect than shape. Thus, areas where letters change but color does not are hard to spot (e.g. p, m, a, and l all have the same color). Because the colors relate to chemical property, it is conceivable that changes of sequence within a color group have a different interpretation in the context of conservation than changes of sequence and color. Even if this is the case, it should be kept in mind that assessing difference boundaries in shapes and colors, many of which appear similar (e.g. g and q, red and purple), is a challenging task. Neither the figure nor the legend motivates the need for using color.

Beauty is the perceptual experience of pleasure or satisfaction. It may arrive through the eye or the mind, and it is the privilege of good scientific communication to do both. Mere form is pretty, but form with function is beautiful.

References

Araldi, D., L. F. Ferrari, and J. D. Levine. 2015. "Repeated mu-opioid exposure induces a novel form of the hyperalgesic priming model for transition to chronic pain." *Journal of Neuroscience* 35(36): pp. 12502–12517.

Cleveland, W. S. and R. McGill. 1985. "Graphical perception and graphical methods for analyzing scientific data." *Science* 229(4716): pp. 828–833.

Ferrari-Toniolo, S., F. Visco-Comandini, O. Papazachariadis, R. Caminiti, and A. Battaglia-Mayer. 2015. "Posterior parietal cortex encoding of dynamic hand force underlying hand-object interaction." *Journal of Neuroscience* 35(31): pp. 10899–10910.

Gehlenborg, N. and B. Wong. 2012. "Networks." *Nature Methods* 9(2): p. 115.

Heer, J. and M. Bostock. 2010. "Crowdsourcing Graphical Perception: Using Mechanical Turk to Assess Visualization Design." In *Proceedings of the 28th International Conference on Human Factors in Computing Systems*, pp. 203–212. Atlanta: ACM.

Johnston, C., N. Campo, M. J. Bergé, P. Polard, and J. P. Claverys. 2014. "*Streptococcus pneumoniae, le transformiste*." *Trends in Microbiology* 22(3): pp. 113–119.

Krzywinski, M. 2013. "Elements of visual style." *Nature Methods* 10(5): p. 371.

Krzywinski, M., I. Birol, S. J. M. Jones, and M. A. Marra. 2011. "Hive plots—rational approach to visualizing networks." *Briefings in Bioinformatics* 13(5): pp. 627–644.

Krzywinski, M. and B. Wong. 2013. "Plotting symbols." *Nature Methods* 10(6): p. 451.

Lex, A., N. Gehlenborg, H. Strobelt, R. Vuillemot, and H. Pfister. 2014. "UpSet: Visualization of intersecting sets." *IEEE Transactions on Visualization and Computer Graphics* 20(12): pp. 1983–1992.

Li, L. L., R. M. Melero-Fernandez de Mera, J. Chen, W. Ba, N. N. Kasri, M. Zhang, and M. J. Courtney. 2015. "Unexpected heterodivalent recruitment of NOS1AP to nNOS reveals multiple sites for pharmacological intervention in neuronal disease models." *Journal of Neuroscience* 35(19): pp. 7349–7364.

Pirsig, R. M. 1974. *Zen and the Art of Motorcycle Maintenance*. New York: Harpertorch.

Procter, J. B., J. Thompson, I. Letunic, C. Creevey, F. Jossinet, and G. J. Barton. 2010. "Visualization of multiple alignments, phylogenies and gene family evolution." *Nature Methods* 7(3): pp. S16–S25.

Strunk Jr., W. 1920. *The Elements of Style*. Ithaca: Priv. print.

Tufte, E. 1992. *Visual Display of Quantitative Information*, Second Edition. Chesire: Graphics Press.

Vetter, P., B. Butterworth, and B. Bahrami. 2008. "Modulating attentional load affects numerosity estimation: Evidence against a pre-attentive subitizing mechanism." *Plos One* 3(9): e3269.

Wender, K. F. and R. Rothkegel. 2000. "Subitizing and its subprocesses." *Psychological Research-Psychologische Forschung* 64(2): pp. 81–92.

Wong, B. 2010a. "Color coding." *Nature Methods* 7(8): p. 573.

Wong, B. 2010b. "Salience." *Nature Methods* 7(10): p. 773.

Wong, B. 2010c. "Gestalt principles (part 1)." *Nature Methods* 7(11): p. 863.

Wong, B. 2010d. "Gestalt principles (part 2)." *Nature Methods* 7(12): p. 941.

Wong, B. 2011a. "Salience to relevance." *Nature Methods* 8(11): p. 889.

Wong, B. 2011b. "Color blindness." *Nature Methods* 8(6): p. 441.

Yantis, S. 2005. "How visual salience wins the battle for awareness." *Nature Neuroscience* 8(8): pp. 975–977.

4 Plant(ing) Aesthetics between Science and Art

Lotte Philipsen

Try looking at Plates 8 and 9 without reading the captions and see whether you can tell which one is art and which one is science. If you find that the two images share a number of visual features, which makes it a delicate matter to determine, it makes perfect sense. Both images represent a flower, and both can be said to possess some degree of visual attractiveness, yet, when analyzed more thoroughly, they express different aesthetic ideals, which directly relates to the fact that they belong to different institutional fields. When experts from such different fields, art theorists on the one hand and scientists from the natural sciences on the other, discuss the roles of art and aesthetics in the sciences, a number of confusions may arise, primarily because the art theorist and the natural scientist tend to define the notions of art and aesthetics differently.

The previous chapter (Chapter 3, this volume) demonstrated how aesthetic choices inevitably play a profound role in the clarity of diagrams that represent scientific data; it accounted for the role of aesthetics *in* science or *for* science. This chapter investigates the difference between aesthetics and beauty in science and in art. By analyzing similarities and differences between the two visually attractive, and seemingly comparable, representations of flowers, this chapter explains why, how, and when two images that both represent a flower exemplify two different conceptions of aesthetics. In-depth comparison of the aesthetic ideals underlying the two images reveals the significant differences between applied aesthetic judgment of taste and pure aesthetic judgment of taste – of which only the former acknowledges the representational dimension of an image. In order to further nuance the different aesthetic ideals at work in the two images, the final part of the chapter discusses a third flower representation that is neither science nor art but a piece of critical design – aesthetically camouflaged as science. By elaborating on and comparing these different examples, the chapter highlights different aspects of aesthetics, their correspondingly different aesthetic ideals, and the role of data representation in these different aesthetic ideals.

Visual Beautification in the Science Community

In order to understand the aesthetic aspects at work in Stefan Eberhard's *Plant reproductive parts* (Figure 4.1, Plate 8), it is useful to know that it was awarded the Wellcome Trust's annual Image Award in 2014. The Wellcome Trust is a major U.K.-based

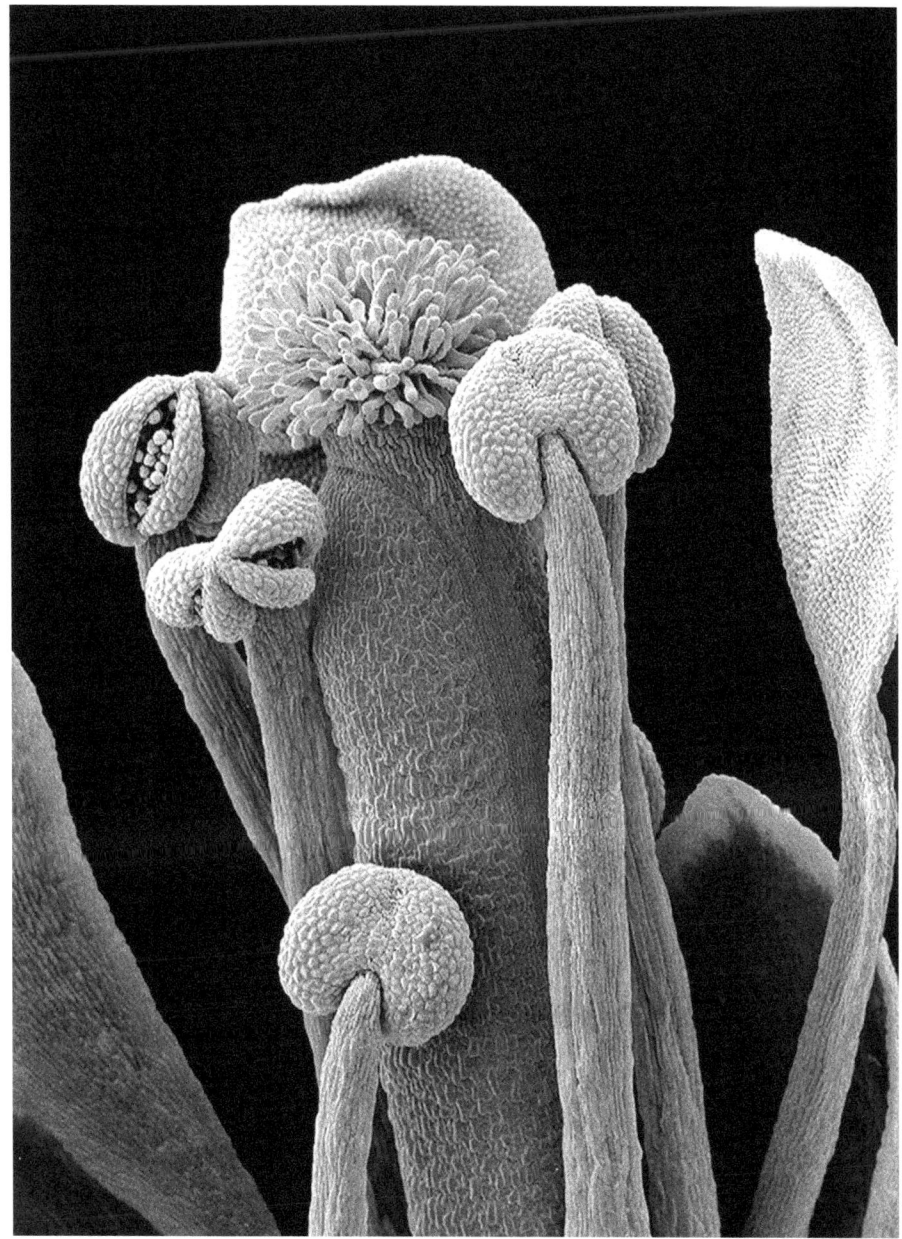

Figure 4.1 (See Color Plate 8) Stefan Eberhard: *Plant reproductive parts*, Wellcome Images 2011. © Creative commons license. 'Scanning electron micrograph (SEM) of an Arabidopsis thaliana flower, also commonly known as thale cress. Some of the anthers are open, revealing pollen grains ready for dispersal. Arabidopsis was the first plant to have its entire genome sequenced and is widely used as a model organism in molecular and plant biology. Horizontal width of image is 1200 μm. Magnification 100x' (Wellcome, 2011).

private research foundation that owns a vast collection of scientific images, the "Well-come Images", and organizes an annual award competition: 'The Wellcome Image Awards are the Wellcome Trust's most eye-catching celebration of science, medicine and life. The Awards recognize the creators of the most informative, striking and technically excellent images that communicate significant aspects of biomedical science' (Wellcome, 2016). The images are firmly positioned in a scientific community, which judges them according to their ability to celebrate and communicate science in an informative manner, but the images' "eye-catching" and "striking" features are equally important, reaching beyond a purely scientific purpose. This double purpose determines the aesthetic ideal governing the image.

Figure 4.1 (Plate 8) depicts the reproductive parts of the *Arabidopsis thaliana* flower. It is a scanning electron micrograph (SEM), which means that, technically, the image is created in the following manner: The flower most likely needs to be dried and/or frozen before it is chemically fixated (Pathan, Bond & Gaskin, 2008) – in the specific case of Eberhard's image, the flower is coated with a nano-thin layer of conductive material (Shearer, 2014). The flower is then placed in a scanning electron microscope, which does not optically see the object but instead detects it by use of an electronic beam that scans the object, measures the electromagnetic emission, and translates that information into a (numeric) set of data containing information on the object's shape and material (Vernon-Parry, 2000). Finally, computer software translates the data into a digital image, which is grayscale – since SEM is not based on optics detecting the wavelength of light, the image has no colors.

The colors in Plate 8 are all added to the original grayscale SEM image subsequently by Eberhard. According to journalist Lee Shearer, Eberhard colors his SEM images 'to allow viewers to separate components better, or because the colors are pleasing to Eberhard's eye, or for some combination of reasons' (Shearer, 2014). Hence, the adding of colors serves two different purposes: One is a diagrammatic purpose in the sense that these artificial colors help scientists as well as laymen to quickly identify the different parts of the flower because each color (however, arbitrarily) corresponds to different parts of the plant that have different functions in the mechanism of reproduction – this diagrammatic purpose is similar to the coloring of the different lines in the London underground map mentioned in Chapter 2 in this volume. Another purpose for coloring is to make the image visually appealing.

The visual ideals governing Eberhard's creation of Figure 4.1 (Plate 8) follow aesthetic preferences that are common across a number of disciplines in the natural sciences. For instance, according to an article on astronomical images, the goal is to create images with 'aesthetic appeal or beauty', which can be obtained by pursuing six guidelines that can roughly be summarized as follows: high photogenic resolution, strong contrast between bright and dark areas, rich colors of a wide variety, the object should fill as much of the field of view as possible, good signal-to-noise ratio, and removal from the image of any trace from image-creating instruments (Christensen, Pierce-Price & Hainaut, 2015). These recommendations specifically concern outreach astronomical images (for concrete examples, see Plates 10 and 12, discussed in Chapter 5, this volume), but obviously they are implicitly at work in Eberhard's image, too.

If the visual preferences behind astronomical images as well as *Plant reproductive parts* are meant to serve a double purpose of helping the viewer to (scientifically and diagrammatically) distinguish between different details in the image *and* be beautiful, the images do not serve both purposes at the same time in the mind of the viewer. When considering the image diagrammatically, the viewer looks *through* the picture plane in order to study what its visual signs represent – that is, the viewer translates the dispersion of different colors on the image's picture plane into the reproductive parts of the *Arabidopsis thaliana* flower. Whereas, when the viewer appreciates (or not) the image's colors, shapes, and composition, the viewer looks *at* the image itself without necessarily considering it a representation of a referent (for elaboration on the sign-referent relation, see Chapter 12, this volume). These two different kinds of looking through or at the image may both appreciate the aesthetics of the image, but they do so according to two different aesthetic ideals: The first involves an applied judgment of taste, and the second, a pure judgment of taste.

Pure and Applied Judgments of Taste

According to philosopher Immanuel Kant (1724–1804), one of the founding fathers of modern aesthetic theory, we pass an aesthetic judgment of taste on phenomena when we claim they are beautiful (Kant, 2005). In other situations, we may pass judgments of reason or moral on (theoretically the same) phenomena, but the question of beauty involves our judgment of taste. Kant distinguished between free and adherent beauty, of which only the former results from a pure judgment of taste, whereas the latter is the result of applied judgment of taste (Kant, 2005, §16). One of Kant's examples of pure judgment of taste seems particularly suitable for the present context:

> Flowers are free natural beauties. Hardly anyone but a botanist knows what sort of a thing a flower ought to be; and even he, though recognizing in the flower the re-productive organ of the plant, pays no regard to this natural purpose *if he is passing judgment on the flower by Taste* [sic](...) A judgment of taste, then, in respect of an object with a definite internal purpose, can only be *pure*, if either the person judging has no concept of this purpose, or else abstracts from it in his judgment.
> (Kant, 2005, §16, italics added)

If a person looks at Figure 4.1 (Plate 8) or at the watercolor painting in Chapter 1 in this volume (Figure 1.4C, Plate 2C) and immediately finds that they are "beautiful visual-izations of the *Arabidopsis thaliana* flower and the bacterium *Streptococcus pneumo-niae*", that person (most likely an expert) categorizes the images under the well-known concepts of that specific flower and bacterium, respectively, and therefore such an aes-thetic judgment is applied. Likewise, if the creators of the images intended to create a beautiful representation of the flower or the bacterium, their aesthetic ideal is one of adhered beauty.

A pure aesthetic judgment is passed if another person (or even the same person in a different situation) looks at the images and finds them unconditionally beautiful without knowing (or paying attention to) what they represent. That person may think that the images are abstract or wonder whether they are perhaps something else but without knowing for sure, which implicitly prompts in the viewer a reflection on the scope and nature of his or her knowledge and sense of perception. This results in an aesthetic experience. Aesthetic judgments are the prerequisite for aesthetic ex-perience, and, according to Kant, they are defined by being 'free' from determining concepts (§16) and 'disinterested' in the sense that they are not guided by specific interests (§2). The only thing that governs aesthetic judgments is taste – the taste of the subject passing the judgment *as if* it had universal validity instead of being a matter of personal preference (Kant, 2005, §19–20). Therefore, it is impossible to establish general rules about what an object or an image should look like in order to prompt aesthetic experience in the viewer. Pure aesthetic judgment and experience simply cannot be guided by determined concepts.

Accordingly, the fact that Eberhard's *Plant reproductive parts* originates from a scientific data representation, a scanning electronic microscopy of an *Arabidopsis thaliana,* does not as such prevent it from being a matter of what Kant terms pure beauty. The image would be a matter of pure beauty if the viewer was to pass a pure aesthetic judgment of taste on it, but as long as the viewer considers the image to be a beautiful *representation* of something (the reproductive parts of *Arabidopsis thaliana*), it is a matter of applied beauty. The fact that it may be a truthful representation of *Arabidopsis thaliana* does not render the image more purely beautiful. According to platonic philosophy, the good and the beautiful were closely related, but modern aesthetic theory (from Kant onward) does not consider the concepts to be automatically dependent on each other.

Visual Beauty in the Art Community

On the face of it, the next example looks as if it shares the same aesthetic ideal as the colorful manipulation of the *Arabidopsis thaliana*. Plate 9 by Robert Mapplethorpe (1946–1989) depicts a poppy flower, and visually this image, too, meets the recommendations given to astronomical images in terms of resolution, contrast, color, composition, noise reduction, and no trace of the apparatus used in the photographic process. Although Eberhard's and Mapplethorpe's images do not look the same, they share the same visual aesthetic. They do not, however, share the same aesthetic ideal.

The difference in aesthetic ideals between the *Poppy* image and Stefan Eberhard's awarded image of an *Arabidopsis thaliana* is not related to their different technical means of production. Whereas Eberhard's image is the result of digital manipulation of an existing grayscale SEM image, *Poppy* is an analogue photograph meaning that the flower has been seen, rather than felt, by the camera in the sense that light beams reflected by the flower have been fixated on photoactive film, which was subsequently processed chemically and finally transferred to photographic paper in a darkroom. Hence, the process of creating an analogue photograph, too, is the result of what some might call manipulation (careful arrangement of light around the flower, adjustment of camera settings, selection of chemical exposure, etc.).

The significant difference between the aesthetics ideals of Eberhard's *Plant reproductive parts* and Mapplethorpe's *Poppy* is that they belong to fundamentally difference discursive fields. The former belongs to the Wellcome Image Collection, and a print of the latter is jointly owned by the Los Angeles County Museum of Art and the Getty Museum and is frequently on display in various exhibitions in leading art museums and galleries worldwide. This means that *Plant reproduction parts* is part of the scientific field, whereas *Poppy* is part of the so-called artworld.

Why is Mapplethorpe's image part of the art world when Eberhard's is not, even though they share the same visual aesthetic? Most people are familiar with at least some of the institutions representing the art world, whereas art theory as such may not be part of laypeople's knowledge. However, art theory plays an important role in explaining why some visually attractive images are not art, while others are.

What Is "Art"?

Obviously, "art" is not a copyrighted term or a protected title, so anyone is free to use the notion as he or she wishes just as anyone is free to use the term "microbiology" to describe whatever he or she may find that term to cover (e.g. bonsai trees). But in both cases, there is a difference between laypeople's use of the terms and methodical correct usage determined by the academic disciplines of art theory/art history and biology, respectively. Every discipline operates with a definition of its objects of research. Academic research in art differs from creation and appreciation of visual art in the sense that it studies how, why, when, where, and for whom art is art.

The definition of art is dynamic in the sense that it changes from one period to another and from one cultural or regional context to another. Throughout history, Western art theory has shifted its focus of attention between different features detectible in the works of art – in order for an object to be "art", it should, for example, mimic the visible world in a specific way, consist of specific materials, praise specific persons/gods/ideals, etc. Across these changing definitions, a common characteristic was that the definitions were descriptive and evaluative at the same time. For example, the eighteenth-century French philosopher of aesthetics Charles Batteux (1713–1780) defined art as that which imitates nature and explained how and why that was good (Batteux, 2012); in 1811, the English philosopher of law Jeremy Bentham (1748–1832) claimed that a defining feature in art was to please people, which would prevent them from starting wars out of boredom (Bentham, 1998); and American art critic Clement Greenberg (1909–1994) in the 1930s–1960s defined modern art as that which focuses on and promotes its own specific medium (in painting: flatness and color), since it would be dishonest for art to imitate something else, such as nature (see, for instance, Greenberg, 1992).

Despite their differences, a common characteristic of the above art theories is that their definitions are not entirely descriptive (what is art?); instead, they are predominantly evaluative (what is *good* art?). An imagined example from biology would be if a biologist's definition of "plant" was informed by his or her ideas about what defines a "good plant" (the one that blooms in the fall, the one that is very big, or the one that spreads to cover an acre at the quickest pace?). In such evaluative definitions, whatever features a specific period considers to be "good" features monopolize the notion of art as such. As a result, there is no scope for distinguishing between "good" and "bad" art.

A radically different art theoretical approach is suggested by the so-called institutional art theory, which caught on in the 1960s and which is solely descriptive. Institutional art theory takes into account the fact that contemporary art is such a diverse field in its use of media/materials and its political/social references that no single criterion in respect of form or content is able to include all works of art. Consequently, institutional art theory detects the defining criteria, not in the works of art themselves, but in the professional, institutional framework that handles the works of art. According to institutional art theory, an object or phenomenon is defined as art if the professional art institutions (museums, art critics, academic departments of art history, etc.) treat it as art (Danto, 1964). This definition is purely descriptive. It works even though one gallery visitor may find a particular work of art fascinating or beautiful, and another visitor may find the same work utterly disgusting, simply because it is a definition of art, not good art or bad art.

Plant reproductive parts or the watercolor painting in Chapter 1 does not comply with institutional art theory's definition of art, because it is not treated as art by the professional art institutional apparatus. It is not added to the collection in an art museum, analyzed or reviewed in established art journals, or on display at professionally curated art exhibitions like *Poppy* is.

The different institutional framework for the two images is not at all accidental since these institutions obviously do not acquire works on a totally random basis. The fact that Stefan Eberhard is a research professional with the Complex Carbohydrate Research Center at the University of Georgia, whereas Robert Mapplethorpe was part of the 1970s and 1980s avant-garde art community in New York, means that their images originate from specific, and very different, institutional contexts and are viewed as phenomena belonging to the discourse of science and art, respectively. It should come as no surprise that scientific institutions focus on images from the science community and that art institutions focus their attention on images from the art community – even if the images sometimes look similar and appear to visually share the same aesthetic ideals. Scientific images and art images are not to be classified hierarchically, in the sense that one type is better than the other. They are simply different kinds of images because of their different institutional framings, which have different aesthetic ideals. Science is about generating knowledge, and aesthetic features may be applied as a means to that end, whereas art is a domain, which, due to its history, ideally gives rise to pure aesthetic judgments of taste and critical reflection but which, according to institutional theory of art, is today a domain governed by its own mechanism of self-sustainment. This does not at all exclude the possibility of works of art giving rise to aesthetic judgments of taste.

Institutional art theory's descriptive rather than evaluative definition of art explains why we are able to pass aesthetic judgments of taste on phenomena that are not art, such as scientific images. A viewer may find an image that is not art appealing, attractive, or beautiful. Likewise, there is no guarantee that all viewers of an institutionally sanctioned work of art may find that particular work appealing or beautiful. However, even if institutional art theory seems to have eradicated normative evaluation criteria from the definition of art, aesthetic judgment of taste still play a very important role in the art world today: First, because after the Enlightenment, the very function of art, its theoretical *raison d'être*, was that it should prompt aesthetic experience instead of being an instrument for other purposes (religious, political, scientific, etc.). Second, ideally and officially, the experts in the art institutions who collect and buy art on a daily basis do so according to their aesthetic judgment of taste – even though in reality a number of pragmatic and financial motives are involved in the processes. Therefore, the aesthetic *ideal* governing the art world, to which *Poppy* belongs, is one of pure judgment of taste.

Pure aesthetic judgment of taste requires that we are not able to conceptually determine what we are witnessing, and, as such, it is not bound to either scientific images or art images. However, both science and art are well-established categories, and very often we know in advance whether the image of a plant we look at belongs to the science or the art domain.

Critical Representation of Science Representation

The Plant Sex Consultancy (Figure 4.2) is an example of a work where no such safe institutional haven exists. This work was created in 2015 by a group of four artists, scientists, and designers (Leopoldseder, Schöpf & Stocker, 2015: 78–79, psx-consultancy.com 2015), and it belongs to a set of practices named in the 1990s as "critical design". The work is staged as a consultancy service that provides help for six different plants that experience sexual difficulties. The consultancy assists each plant by designing tiny 3D-printed devices that help the plant spark its reproductive activity. *The Plant Sex Consultancy* has a web page and a printed booklet, which present the consultancy's services, and it is presented to viewers in exhibitions of specimens of the six different plants with their attached aid devices in the gallery along with text that for each of the six plants offers 'Plant Statement', 'The Problem', and 'The solution'. The *Curcuma alismatifolia* flower's consultation is, for example, described like this:

> Plant Statement: "Usually I'm not considered as a growing plant but a cut flower or a spice. I sleep from November to May every year in the form of a rhizome bulb where all my belongings are stored." The Problem: "I come to you because I literally have NO SEX LIFE. I'm infertile... My offspring are just clones, and I hate looking at younger versions of myself, blossoming in my vicinity." The solution: The rhizome is dug out when the plant goes into dormancy. A weather balloon is then tied to the rhizome, carrying it to the edge of the Earth's atmosphere, where the rhizome receives a huge amount of radiation and thus mutates the genome. As the balloon rises closer to the space, the temperature and pressure burst the balloon and the rhizome is dropped back to Earth far from where it originated, planting a mutant different from its clonal lineage.
>
> (Lin, Stamatis, Weiss & Petrič, 2014, original capitals)

As seen in Figure 4.2, the drawing that accompanies this specific "consultation" displays the suggested small 3D-printed balloon that, according to the "solution", is to be attached to the plant to help it rise into the atmosphere where radiation will prompt it to mutate.

Figure 4.2 and the consultation text demonstrate how *The Plant Sex Consultancy* closely resembles real scientific experiments and research. Nothing about the idea of attaching small, helium-filled balloons to plant parts that lift them into the atmosphere for mutation caused by radiation, and the subsequent bursting of the balloon resulting in the return of the mutated plant to Earth, is un-scientific as such. Every step in the procedure is part of an overall rational plan of promoting plant diversity. So why is *The Plant Sex Consultancy* not science, when it clearly makes use of scientific research?

The work offers a clumsy solution to the specific problem by interpreting the practice of synthetic design too literally. Synthetic design is a widely used practice where humans modify (artificial or natural) organisms on a molecular, genetic level – that is, the

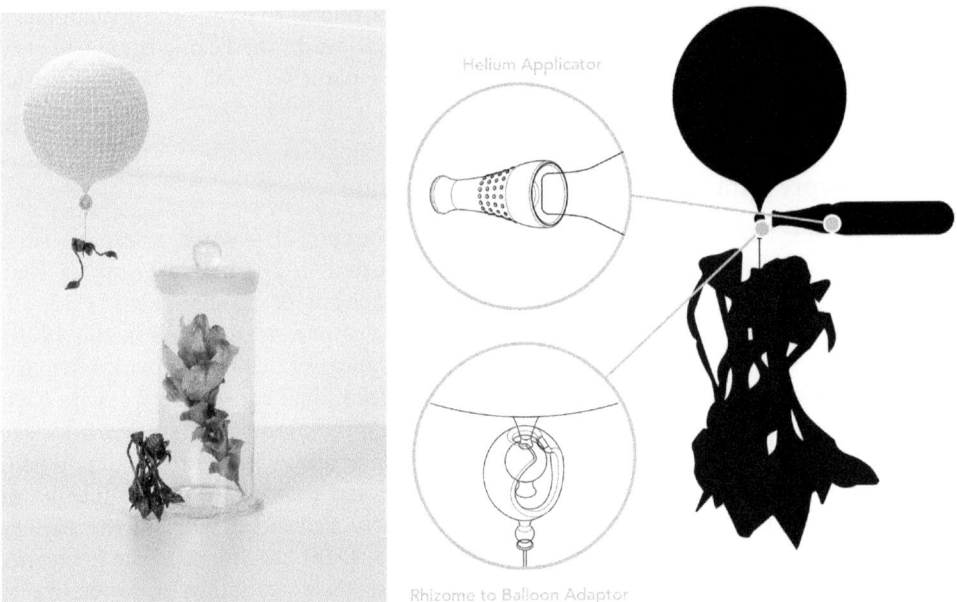

Figure 4.2 "Curcuma alismatifolia" from *The Plant Sex Consultancy*, 2014 by Pei-Ying Lin, Dimitris Stamatis, Jasmina Weiss, and Špela Petrič. Used by permission.

organisms are designed synthetically instead of evolving through natural selection and mutation. But *The Plant Sex Consultancy* offers modifications and add-ons not in the form of genetic manipulation as such but in the form of 3D-printed plastic devices that are physically applied to the individual specimen. In other words: The work, oddly, offers a hardware solution (physical applications) to a software problem (the genetic code of the plant). If the aim of the work was to actually create mutations of a rhizome, it would be much easier to synthetically re-design the plant by gene manipulation in the lab than to go through the trouble of lifting the plant halfway into outer space, hoping that radiation would do the job, and that the mutation would subsequently find its way back to a spot of soil from where it would grow and spread. Still, even "poor" science is science insofar as it is judged according to scientific standards. If, however, the aim is to leave the audience genuinely puzzled and undetermined as to the status of what *The Plant Sex Consultancy* is (therapy, design solutions, scientific experiment, a joke, or something yet different?) then the suggested solution does a much better job.

The work confuses its viewers by appropriating and re-mixing visual as well as linguistic features from domains that normally have nothing in common. Linguistically, the work mixes rhetoric known from agony columns with that from biotech industry: The work applies the linguistic approach of therapeutic agony columns by using the rhetoric of a first-person speaker that reveals personal problems ('I come to you because I literally have NO SEX LIFE...') and by staging itself as a consultancy service. Conversely, the use of Latin plant names provides the work with a scientific appearance, which is stressed by the fact that the *The Plant Sex Consultancy*'s website includes an extended list of references to real scientific publications (see Lin, Stamatis, Weiss & Petrič, 2014).

The rhetoric of articulating a "problem" and a "solution" is well established in both domains: In agony columns, the expert/consultant suggests solutions to frustrated advice seekers, and in the natural sciences the problem-solution rhetoric is used as an effective way of communicating complex matters in outreach material to the public or in grant applications to foundations. Hence, in *The Plant Sex Consultancy*, the problem-solution scheme works as a semiotic shifter – a sign whose meaning changes depending on the context created by adjacent signs (see also Chapter 12, this volume) – because the context is ambiguous insofar as the work simultaneously articulates itself as scientific research and personal consultation. The problem-solution rhetoric enables *The Plant Sex Consultancy* to clearly and effectively communicate the specific scientific issues concerning each plant in an accessible manner to the broader public, while at the same time it generates a great extent of uncertainty in the viewer/reader about the seriousness and truth of the scientific content due to the agony columns context. Non-biologists may find themselves in doubt as to whether the matter presented (e.g. that *curcuma alismatifolia* is "infertile") is in fact presented as a joke or if all the information in the work is made up. Or maybe this is actually all serious. Maybe *The Plant Sex Consultancy* is a private company in agricultural or biotechnological industry that offers its services to other companies with a need for optimizing the reproducibility of specific plants. But what, then, about the implied references to sex toys, suggested by the plastic ad-ons? In addition to the advanced suggested solutions to the plants' problems, the work's scientific appearance is promoted by its visual design, which resembles that of the biotech industry, too.

The image design of *The Plant Sex Consultancy* effortlessly combines different visual typologies like photographs of real plant details (like the flower in the glass jar); computer-generated visualizations of the plastic devices (the helium balloon); and graphic illustrations that clarify design details by enlarged sketches located on a black plant-with-device silhouette. White backgrounds create the sense of everything existing in an ideal, pure, virtual space that resembles a visual promise of a sterile future untainted by the problems of everyday reality. As such, *The Plant Sex Consultancy* taps into the visual rhetoric design of research and innovation departments in companies like Pfizer or GlaxoSmithKline, but it does so from the outside of the biotech and science discourse, since institutionally it is positioned elsewhere. It was developed at the Ljubljana Design Biennial 2014 (Fredrickson, 2014) and exhibited at the Ars Electronica Festival 2015, where it received honorary mention (Leopoldseder, Schöpf & Stocker, 2015: 78–79). As such, it is not science but neither is it art, and design-wise it lacks efficient functionality.

The Plant Sex Consultancy thus establishes a room for critical reflection rather than for concrete biological problem solution, even if mimicking the latter is the crucial means for the work's ability to obtain the former. As such, the project functions as critical design – a kind of aesthetic practice, which, according to designers Anthony Dunne and Fiona Raby who coined the term, is characterized in the following manner:

> Critical design uses speculative design proposals to challenge narrow assumptions, preconceptions, and givens about the role products play in everyday life. […] On the most basic level it is about questioning underlying assumptions in design itself, on the next level it is directed at the technology industry and its market-driven limitations, and beyond that, general social theory, politics, and ideology.
>
> (Dunne & Raby, 2013, 34–35)

The point in mimicking the aesthetic means (the visuals and the linguistic rhetoric) by which the biotech industry represents its facts and data is that it allows for critical reflection on these specific means of representation. In other words, it enables the audience to see and read the industry's means of communication as representational acts and not merely as facts (for elaboration on representational acts, refer to Chapter 12, this volume). Very often art imitates the aesthetic of other domains and questions underlying assumptions of everyday life and ideology, so what is the difference between art and critical design? According to Dunne and Raby:

> Critical design might borrow heavily from art's methods and approaches but that is it. We expect art to be shocking and extreme. Critical design needs to be closer to the everyday; that's where its power to disturb lies.[...] If it is labeled as art it is easier to deal with but if it remains design, it is more disturbing; it suggests that the everyday life as we know it could be different, that things could change.
>
> (Dunne & Raby, 2013, 43)

The Plant Sex Consultancy does not work as a representation of scientific data, but it offers a representation of the aesthetic features of scientific communication and data representation. And it does so by using an aesthetic articulation that potentially causes a pure aesthetic judgment of taste in the viewer of the work. In fact, ideally, it causes an aesthetic judgment of taste, because otherwise it would fail as critical design, and it already fails as science (institutionally, it does not belong in the science domain) and as art (neither does it belong in the art domain), which is the whole point: The un-determinacy of critical design, in this case its ambiguous position between science and art, provides it with profound aesthetic potential. The chapter's three different representations of plant are aesthetically very different – not because they look different, but because they express different aesthetic ideals due to their different institutional framework of existence, which directs how we perceive and comprehend them.

References

Batteux, C. 2012. "The Fine Arts Reduced to a Single Principle". In *The Bloomsbury Anthology of Aesthetics*, edited by J. Tanke and C. McQuillan, pp. 140–157. London: Bloomsbury Publishing.

Bentham, J. 1998. "Reward Applied to Art and Science". In *Art in Theory 1815–100*, edited by C. Harrison, P. Wood, and J. Gaiger, pp. 149–151. Oxford: Blackwell Publishers.

Christensen, L. L., D. Pierce-Price, and O. Hainaut. 2015. "Determining the aesthetic appeal of astronomical images". *Leonardo* 48(1): pp. 70–71.

Danto, A. 1964. "The artworld". *The Journal of Philosophy* 61(19): pp. 571–584.

Dunne, A. and F. Raby. 2013. *Speculative Everything: Design, Fiction, and Social Dreaming*. Cambridge: The MIT Press.

Fredrickson, T. 2014. PSX consultancy designs sex toys for plants at BIO 50, Original Publication. www.designboom.com/http://www.designboom.com/design/psx-consultancy-plant-sex-toys-bio-50-10-06-2014/. Accessed 26 Nov 2016

Greenberg, C. 1992. "Toward a Newer Laocoon". In *Art in Theory 1900–1990*, edited by C. Harrison and P. Wood, pp. 554–560. Oxford: Blackwell Publishers.

Kant, I. 2005. *Critique of Judgment* (German edition 1790). Mineola: Dover Publications.

Leopoldseder, H., C. Schöpf, and G. Stocker. 2015. *CyberArts 2015: International Compendium—Prix Ars Electronica*. Ostfildern: Hatje Cantz.

Lin, P. Y., D. Stamatis, J. Weiss, and Š. Petrič. 2014. The Plant Sex Consultancy—Giving small pleasures to the planted. Original Publication, http://psx-consultancy.com/. Accessed 15 Dec 2015.

Pathan, A. K., J. Bond, and R. E. Gaskin. 2008. "Sample preparation for scanning electron microscopy of plant surfaces—Horses for courses". *Micron* 39(8): pp. 1049–1061.

psx-consultancy.com. 2015. http://psx-consultancy.com/. Original Publication, http://psx-consultancy.com/. Accessed 28 April 2016.

Shearer, L. 2014. *UGA's Stefan Eberhard Photographs Microscopic Worlds*. Georgia: Athens Banner-Herald.

Vernon-Parry, K. D. 2000. "Scanning electron microscopy: An introduction". *III-Vs Review* 13(4): pp. 40–44.

Wellcome. 2011. Plant reproductive parts. Original Publication, www.wellcomeimageawards.org/2014/plant-reproductive-parts. Accessed 19 Nov 2016.

Wellcome. 2016. Wellcome Image Awards. Original Publication, www.wellcomeimageawards.org/about/about-the-awards/. Accessed 1 Oct 2016.

5   Visualizing the Invisible Universe

Steen Hannestad

This chapter describes how astrophysicists look at and visualize outer space – something that is very big, very far away, and, in most cases, has ceased to exist millions of years ago. The chapter accounts for the rising complexity in astrophysics visualization: from the challenges of taking photographs of outer space phenomena that emit light in wavelengths not detectable by the human eye and therefore in need of a lot of manipulation in the visualization process, to the methods of visualizing phenomena that are in their very essence invisible since they consist only in mathematical data.

This chapter also discusses the reasons for visualizing outer space, considering the fact that the data is, in reality, invisible. The importance of communicating science is one reason. Another reason is that the visualizations are able to provide different scientific insights than pure mathematics. Often the most difficult part of research is to understand which questions to ask, and in this context visualization is extremely useful because it is much closer to human intuition than formal equations. In many cases, astrophysical data must be studied visually before one can begin to describe it mathematically.

The Birth of Astrophysical Visualization

Visualization of the universe dates back to prehistoric times – several depictions of stellar constellations are known from cave paintings, one example being the visualization of the Pleiades star cluster in a painting from the Lascaux cave in France (see e.g. Kaulins, 2003). However, until the invention of the telescope early in the seventeenth century, visualization of astronomical objects was restricted to depictions of what an astronomer could observe with the naked eye.

This changed radically when Galileo as the first in history pointed a telescope toward the skies and used it to, among other things, discover the four large moons of Jupiter. During the following centuries, astronomical visualization increased in complexity. However, it is still limited by the fact that, with the exception of a few cases, the brightness of astronomical objects is so low that it is impossible for the eye to rely on its cone photoreceptors – the ones we use for color vision. Instead, the eye must rely on the rod receptors, which means that astronomical objects are always seen in a greenish hue with no color differentiation.

The brightly colored images of interstellar gas clouds and galaxies, which most people have seen in newspapers or magazines, only became possible with the invention of astrophotography in the mid-nineteenth century. This allowed for long-time exposure of images, and the measurement of distinct colors became possible.

Figure 5.1 (See Color Plate 10) The Orion nebula shown in a combination of infrared, visible, and ultraviolet light. The green hue stems from hydrogen and sulphur in the cloud, whereas the orange colors stem from complex organic molecules.
Image credit: NASA/JPL-Caltech/STScI.

The distinct colors seen in many astrophysical objects are related to the presence of specific elements such as hydrogen, oxygen, and sulphur (Figure 5.1, Plate 10). Today, the detection of elements in astrophysical objects is done via so-called spectroscopy, where the wavelengths of the emitted light are studied in detail (Massey & Hanson, 2011). Historically, visualization via color astrophotography was truly important from a scientific point of view because the colors could never have been seen by the naked eye.

The use of augmented vision through photography was also developed further through the twentieth century, particularly allowing for images to be taken in wavelengths outside the range accessible to the human eye. For example, the sun emits most of its light in the visible spectrum, which is why the human eye is designed to be sensitive to exactly this wavelength range. However, a huge amount of insight can be gained from studying the sun in other wavelengths, ranging from extremely long wavelength radio waves to ultra-short wavelength X-rays. For instance, the solar magnetic field can be studied through X-ray imaging of the sun even though the only trace of the magnetic field of the spectrum visible to the human eye is the presence of sunspots (Figure 5.2).

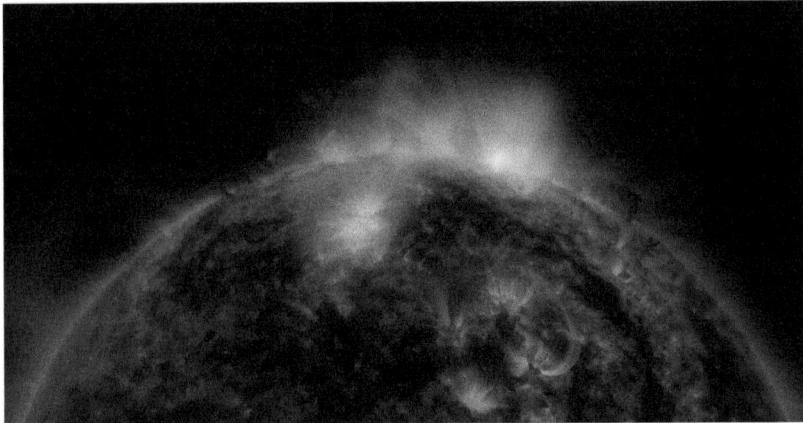

Figure 5.2 The sun in a combination of ultraviolet and X-ray light. The regions of high activity
show the magnetic field outside the solar surface. These large magnetic lobes are
where the solar wind discussed in Chapter 7, this volume, originates.
Image credit: NuSTAR/SDO/MASA.

Modern Astrophysical Visualization

Advances in modern technology have had a decisive influence in astrophysical
visualization. For example, it has become possible to perform detailed studies of the
so-called cosmic microwave background (CMB) radiation, which originates from
the very early universe, only 400,000 years after the Big Bang (Dodelson, 2002).
This manifests a fundamental property of light, namely that it does not travel at
infinite speed. Even though the speed of light, at 186,000 miles per second, is very
fast compared to anything we encounter on Earth, the distances in space are so vast
that the finiteness of the speed of light becomes an important factor when performing
these measurements.

When we look at the sun, we actually see light that was emitted a little over eight
minutes ago. Of course, we know from experience that the sun hardly changes
appearance within such a short time span. However, looking at more distant objects,
this can change. Our nearest neighboring galaxy, the Andromeda Galaxy, lies at a
distance of more than 2 million light-years. Even though a star like the sun lives for
more than 10 billion years, this is not the case for all stars. In fact, some stars have
total life spans shorter than the time it takes for light to traverse the distance between
the Andromeda Galaxy and Earth. In 2 million years, the Andromeda Galaxy does
change its appearance to some extent, and it is literally impossible for us to visualize the
Andromeda Galaxy as it looks right now – we are forever restricted to imaging the past.

This issue becomes intensified as we look at more and more distant objects. In
many cases, we can now study galaxies, which are billions of light-years away,
meaning that we see them as they looked billions of years ago. The finite speed of
light is in many ways both a blessing and a curse. Even though we will never be able
to see astronomical objects as they look this instant, we instead have access to a time
machine. We can actually study the way the universe looked billions of years ago and
thus study objects, which most likely ceased to exist long ago.

When light propagates through the expanding universe, its wavelength increases with the expansion of the universe. This effect can be seen as the so-called red-shifting of light, referring to the red end of the wavelength spectrum. For objects in the nearby universe, this effect is not dramatic and could not possibly be seen with the naked eye. This is very different for more distant objects. The light from a galaxy billions of light-years away has typically been shifted into what is the infrared range and is thus invisible to the naked eye. Even with the largest telescope, we would only be staring at a seemingly empty universe beyond a certain distance. The next generation of space telescope – the James Webb telescope due to be launched in 2019 (www.jwst.nasa. gov) – will not contain equipment to measure light in the visible spectrum. Instead it will focus completely on the infrared range, which is more interesting for studies of the very distant universe as well as planets orbiting other stars.

The CMB is the most extreme example we have of the red-shifting phenomenon. It is a remnant of the very early universe and was emitted when the temperature of the entire universe was comparable to that on the surface of the sun. The CMB initially consisted of light in the infrared and visible parts of the spectrum and would have looked as bright as the sun in all directions. However, by now the radiation has cooled to a faint afterglow of what it once was. Its intensity is no more than a trillionth of what it was when it was emitted, and the wavelength has now been shifted to the microwave part of the spectrum, completely invisible to the naked eye.

Observations of the CMB from Earth are quite difficult because microwaves are efficiently absorbed by the water in the atmosphere (the same effect used to heat food in a microwave oven), and observations are best done by satellites. The measurements done by such satellites have provided exquisite images of the CMB, in particular the European Planck satellite has provided us with an amazing amount of information about small intensity variations of the CMB across the sky (Figure 5.3, Plate 11). The

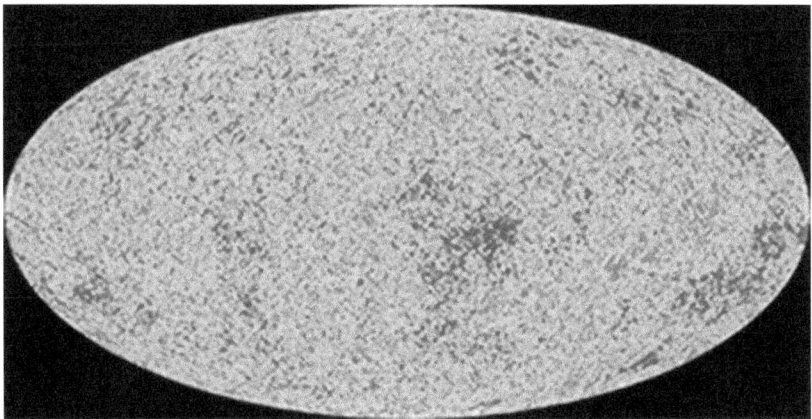

Figure 5.3 (See Color Plate 11) The cosmic microwave background (CMB) as detected by the Planck satellite. The color-coding reflects the intensity of the radiation in various directions on the sky, with red signifying high intensity and blue corresponding to low intensity. The difference in intensity between the red and the blue regions is only about 1 part in 100,000.

Image credit: ESA and the Planck Collaboration.

relative differences between bright and dim regions in the sky are only 1 in 100,000 and would be completely invisible when visualized without augmentation of the differences. However, these small variations are crucial to our understanding of the universe. They arise from small variations in the density where the light was originally emitted. These are the small density variations, which over time and through the effects of gravity grow to become the stars and galaxies in the present-day universe. Without these small fluctuations, we would not be here to worry about them.

Even though the CMB can be studied in great detail without the need for actual visualization, there is a lot to be gained from images such as the one from the Planck satellite (Figure 5.3, Plate 11). It is evident that different regions of the sky look roughly similar on average. This is important information and indeed predicted by our current physical models of the universe.

As an example of how astrophysical data can be processed to provide other types of visualization, the CMB measurements from the Planck satellite can be processed through a so-called angular harmonic transform to give the angular power spectrum of the CMB (Figure 5.4). The angular power spectrum provides a powerful way of visualizing the data through a diagram. The peak located at an angular scale of around 1° tells us that bright and dim regions (i.e. red and blue pixels in Figure 5.3, Plate 11) on the sky are on average separated by about 1° on the sky. This provides us with information otherwise impossible to extract by any human directly from the image of the CMB, but through extensive processing of the original image of the sky in microwave radiation, we can gain a huge amount of insights. Thus, the diagram in Figure 5.4 represents data that is invisible in – yet stems from – the image in Figure 5.3 (Plate 11).

The CMB is the earliest image of the universe we can ever take. One could think that by looking further and further away we would eventually be able to see all the way back to the Big Bang. After all, that seems like the natural limit to this process. However, as we go back in time, the universe was hotter and denser than it currently

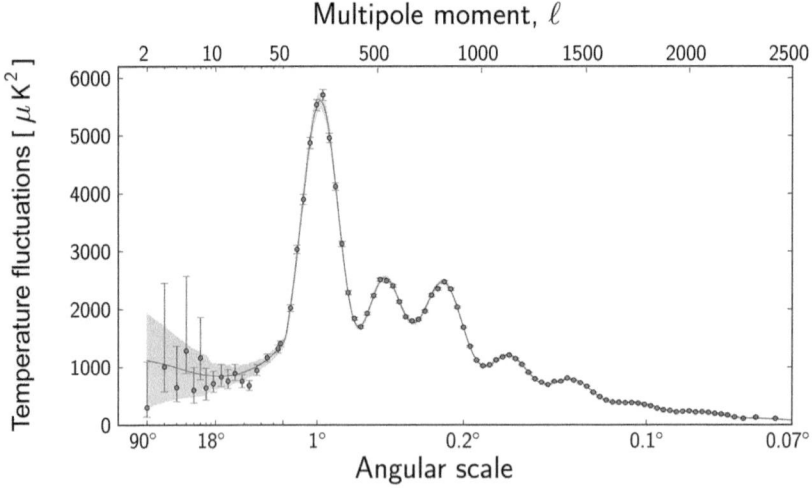

Figure 5.4 The angular power spectrum of the CMB as measured by the Planck satellite.
Image credit: ESA and the Planck Collaboration.

is. At the time when the CMB was formed, the density was such that the universe was only just transparent to light. Earlier in its evolution, the universe was opaque, and any light emitted would have been immediately scattered or reabsorbed. The images of the CMB taken by the Planck satellite, like Figure 5.3 (Plate 11), are actual images, even though they are of course highly manipulated, but we can never take images of the universe prior to the formation of the CMB, because we can never capture light from that period. That does not mean we cannot visualize the universe prior to this point, but in that case, the data used for doing so would have to be numeric, and the resulting images would be based solely on computer models.

The Truly Invisible

In many cases, image processing and augmentation fall short. In fact, we now know that most of the universe consists of matter and energy that is truly invisible in the sense that it neither emits nor absorbs light of any kind. Most of the mass in our galaxy consists of the so-called dark matter, an, as yet, unidentified type of matter, which interacts with our "normal" world only through gravity ("normal" in this context means anything consisting of known elementary particles, including therefore what one would normally think of as the entire universe). In total, there is at least five times as much dark matter as normal matter in the universe, and life in the universe would be impossible without dark matter – still, we do not yet know exactly what the dark matter is. Perhaps the most likely candidate is a new type of stable and very heavy elementary particle, and there is hope that the first physical traces of such a particle will be detected or "seen" in the high-energy particle collisions in the Large Hadron Collider at CERN. In rare cases, these particles might be produced in high-energy particle collisions, and even though they cannot be directly measured in particle detectors, they can be indirectly seen. The fundamental principle of energy conservation applies in particle physics, and therefore any particle not measured in the detector will show up as an apparent violation of energy conservation. Since it is clearly impossible to understand how the universe behaves without understanding the dark matter, one is immediately faced with the question of how to probe dark matter and subsequently how to visualize something that is truly invisible.

In the case of dark matter, we must rely on the one point where it can actually be "seen", namely through its gravitational influence on other matter. For example, we can measure the speed with which stars rotate around the center of the Milky Way. In order to maintain the roughly circular orbits of stars, a certain amount of mass is needed, and this mass depends on the total mass contained in the galaxy. This information is also used in our own solar system where the mass of the sun can be measured by measuring the rotation velocity of planets as a function of their distance from the sun. By measuring the rotation velocity of many stars in our galaxy, we can make a rough map of the density of dark matter. This map can be visualized, even though it is in no way an actual image but rather based on a computer model generated through the measurement of completely different objects, namely stars. The type of density map derived from these observations is quite similar in nature to the density and pressure maps of the atmosphere used in weather forecasting and climate studies, which also rely on graphic visualization of physical parameters, which are not directly visible. Extensive observations of many galaxies have revealed that dark matter is distributed very differently from the stars in galaxies. A galaxy typically consists of a disk-shaped distribution of stars embedded in a much larger and almost spherical "halo" of dark matter (e.g. Dodelson, 2002).

This method of measuring the presence of dark matter in galaxies is also used on larger scales. Most galaxies are contained in big galaxy clusters, often consisting of a thousand or more individual galaxies. In this case, it is no longer possible to measure the velocities of stars in order to probe the density of dark matter. Instead it is possible to get a rough idea of how much dark matter is contained within a cluster by measuring the velocities of galaxies in the cluster. In fact, this is how dark matter was first discovered in the 1930s by the Swiss-American astronomer Fritz Zwicky. Zwicky realized that the velocities of galaxies in the Coma Galaxy cluster were much higher than anticipated. Typical velocities were such that they seemed to be much higher than the required velocity to escape the cluster. In other words, it looked as if the cluster should not really have been a bound object but rather an accidental collection of galaxies. The universe contains many galaxy clusters, of which we know of thousands, and the probability that they are all coincidental is practically zero. Instead there must be some physical explanation for the abundance of these objects. Zwicky almost immediately concluded that galaxy clusters must be gravitationally bound objects because they contain far more mass than what is in the individual galaxies.

These early observations of galaxy clusters have since been confirmed in many studies, and we now know that typical galaxy clusters contain about five times more dark matter than normal matter (Dodelson, 2002). However, the measurement of velocities of galaxies in clusters remains a relatively crude and vague method for probing the dark matter. Within the last two decades, a novel and much more precise method has been used with great success, relying on the phenomenon of gravitational lensing.

We are used to the fact that light can be distorted by propagating through transparent objects like glass or water, or through reflection in non-planar mirrors. However, we normally think of light as propagating in straight lines through empty space. Fermat's principle – named after the French mathematician Pierre de Fermat (1601–1665) – in optics seems to tell us exactly that: Light always takes the shortest possible path, and in empty space that surely ought to be a straight line.

However, Albert Einstein's theory of general relativity describes gravity as a bending and curving of space-time. When the Earth revolves around the sun, it happens because the mass of the sun distorts the local space-time. The Earth actually moves in a straight line, but the space-time distortion means that the line is no longer straight but rather takes the shape of an almost circular orbit around the sun.

As for light, the distortion of space-time also means that even though light always obeys Fermat's principle, it does not move in exact straight lines. The effect can be measured in our own solar system: Stars located very close to the position of the sun in the sky seem to shift their position by a small amount relative to where they are when the sun is far away. This is not caused by any actual shift in position but rather due to the fact that light from the stars is bent by the space-time distortion generated by the sun. In the case of the sun, the effect is minute and measurable only with sensitive instruments, but it can be much more dramatic. For example, it is possible for light to orbit a black hole in a perfect circle.

When galaxies located behind a galaxy cluster are observed, the huge mass of the galaxy cluster distorts space-time to a degree where several images of the same galaxy can appear in different locations. The phenomenon where the shape of galaxies is distorted is known as gravitational lensing and is visually similar to some of the effects seen in optical lensing – when light passes through the galaxy cluster, it is distorted in a way that looks superficially similar to looking at light through water. By measuring thousands of galaxies behind galaxy clusters, it is possible to provide detailed maps of the dark matter in galaxy clusters.

One of the most interesting examples of how this technique can be applied to create astrophysical images is visualization of the so-called bullet cluster, which consists of two very large galaxy clusters that collided several million years ago (Markevitch et al., 2002). When this happens, the gas contained in the clusters is heated by interactions and is slowed down. This is seen as the pink regions in Figure 5.5 (Plate 12). The color is generated by manipulation of the image because the gas actually emits light mainly in the X-ray part of the spectrum. However, the dark matter can only interact through gravity, and therefore the dark matter contained in the clusters is typically affected very little by the collision. By means of gravitational lensing, it has been possible to construct a map of the dark matter in both clusters. This is shown as the blue regions in Figure 5.5 (Plate 12). Evidently, the gas has been slowed down relative to the dark matter in both clusters.

Even though dark matter does not emit or absorb any kind of light, it is not truly invisible because gravity also affects how light propagates. Visualization of our universe has evolved through centuries from simple depictions of what the naked eye can see into advanced visualizations, not just through image manipulation but also to the point of making the seemingly invisible visible. This has had a profound impact on the way we understand the universe. From visualizations of how dark matter must

Figure 5.5 (See Color Plate 12) The Bullet Cluster measured in the visible spectrum and then printed as a collage with additional information measured in X-rays (pink) and through gravitational lensing (blue).

Image credit: NASA/CXC/CfA/Markevitch et al.

be distributed in galaxies and on larger scales, it has become possible to understand many of the properties needed to describe its behavior mathematically. Future use of this interplay between visualizations and formal mathematical calculations will lead to an even deeper understanding of what dark matter is.

References

Dodelson, S. 2002. *Modern Cosmology.* Cambridge: Academic Press.

ESA. 2013, October 18. "A Portrait of the Cosmos as a Young Universe." Cited 2016 September 13. http://sci.esa.int/planck/53103-planck-cosmology/. Accessed 27 April 2017.

Kaulins, A. 2003. *Stars, Stones, and Scholars—the Decipherment of the Megaliths.* Bloomington: Trafford Publishing.

Markevitch, M., A. H. Gonzalez, L. David, A. Vikhlinin, S. Murray, W. Forman, C. Jones, and W. Tucker. 2002. "A textbook example of a bow shock in the merging galaxy 1E 0657 56." *Astrophysical Journal Letters* 567: p. L27.

Massey, P. and M. M. Hanson. 2011. "Astronomical Spectroscopy." In *Planets, Stars, and Stellar Systems,* Vol. 2. New York: Springer.

NASA. n.d. "Explore James Webb Space Telescope." Cited 2016 September 13. www.jwst.nasa.gov. Accessed 27 April 2017.

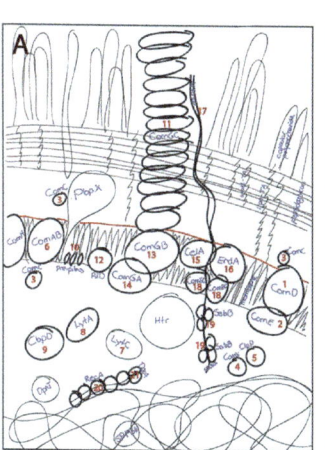
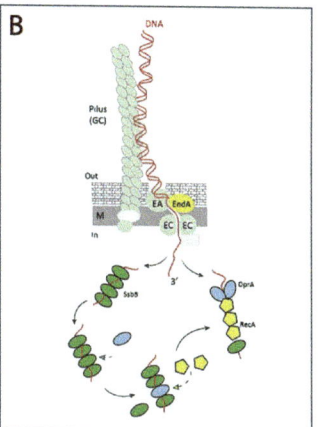
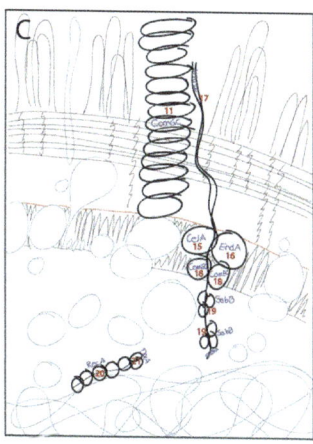

Plate 1 (Figure 1.1) **A.** Rough working sketch of a section of a pneumococcal cell showing the uptake and integration of foreign DNA, known as transformation. Structures with bold outlines and red numbers are described in the short painting narrative in the box. The red line on the surface of the cell membrane marks the boundary between inside and outside. Structures above the red line are surface structures outside the pneumococcal cell, whereas the proteins and DNA shown below the red line are located inside the cell. The numbers in the sketch refer to numbers in the brackets in the text. This rough sketch serves as an overview and as a guide for drawing the final sketch for the painting (see Figure 1.4). **B.** This figure shows how DNA (red) binds the transformation pilus before entry proteins (labeled EA and EndA) process it. The DNA is imported through a channel in the membrane (labeled EC) and bound by protective dark green proteins and integrated in the genome by the blue and yellow proteins. The figure is reprinted with permission from the review *Streptococcus pneumoniae, le transformiste* by Johnston et al. (2014). **C.** Modified version of A, highlighting the proteins shown in both A and B. The comparison between B and C exemplifies how we used traditional journal article illustrations in composing our rough sketch. Browsing through illustrations in the literature, we quickly identified which important proteins to illustrate. Using the proteins that were illustrated most often, we broadened the literature reading to be able to compose the full narrative.

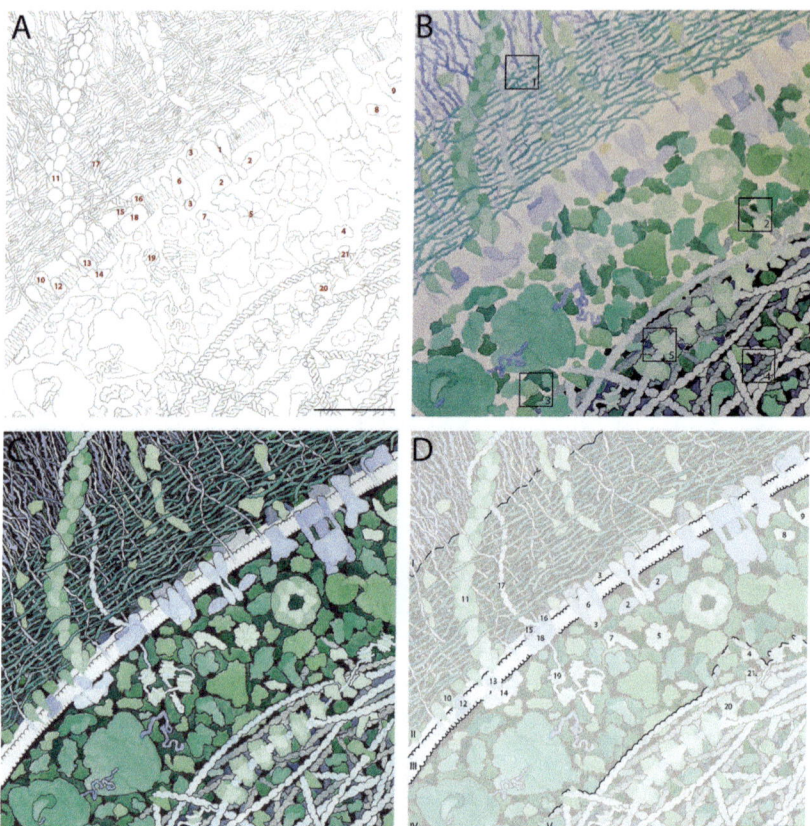

Plate 2 (Figure 1.4) Painting pneumococcal transformation. **A.** Final painting sketch showing the foreground of the painting. The proteins and DNA described in the painting narrative (Box 1.1) are labeled accordingly in red. The black scale bar to the lower right represents 5 centimeters in the painting and corresponds to 25 nanometers in the cell. **B.** This illustration shows the process of the painting. First, the foreground is painted (focus 1). The proteins labeled with red in A are highlighted by using a very light tint of green and blue (focus 2). Darker proteins are added in the middle ground (focus 3), and even darker proteins or DNA, as shown here, are added in the background (4). Black contours are used to outline complexes where two or more proteins or DNA interact (5). **C.** Final pneumococcal transformation painting. Photo by Michael Grøn, Billedmageren – Grøn og Grønborg Fotografi. **D.** Visual explanation for the final painting. The black numbers are identical to the numbering in A indicating the localization of key proteins. From the top left toward the lower right, the painting shows the capsule (labeled I) and the cell wall (II) on the surface of the cell. The cell membrane (III) encloses the interior of the cell, which consists of the cytoplasm (IV) and the nucleoid region (V). Photo by Michael Grøn, Billedmageren – Grøn og Grønborg Fotografi.

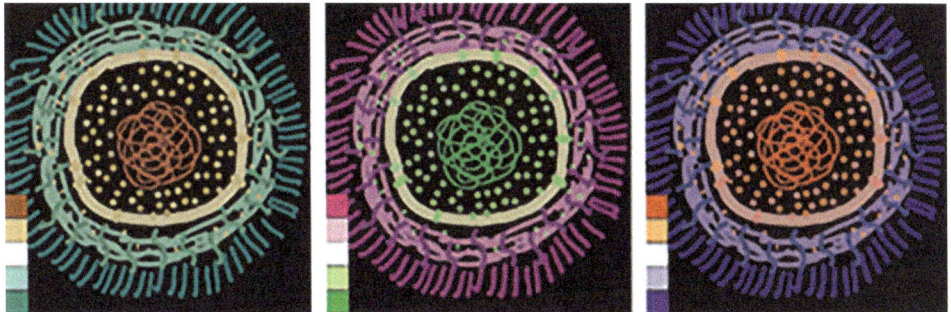

Plate 3 (Figure 1.5) Choosing colors for the painting series. Based on the diverging color-blind-safe color palettes from ColorBrewer.org, we selected the three schemes most pleasing to us as indicated by the color bars in the lower left corner of each illustration. The three illustrations shown here served as a basis for which to decide on a general color palette for the eight future paintings.

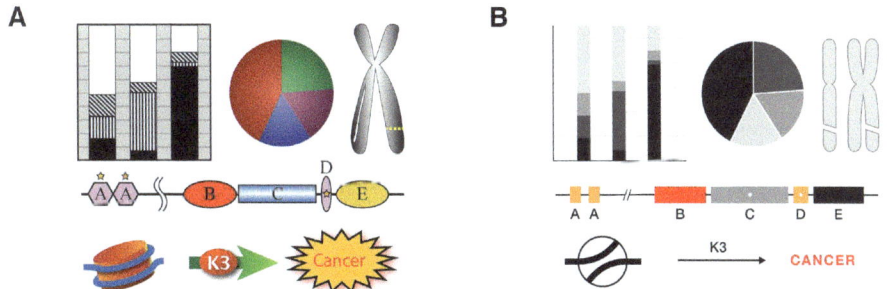

Plate 4 (Figure 3.2) **A.** Florid visuals are distracting and make adding emphasis difficult. **B.** Conservative formatting of shapes from Figure 3.2A more effectively present spatial and semantic relationships between elements and make adding emphasis easier.

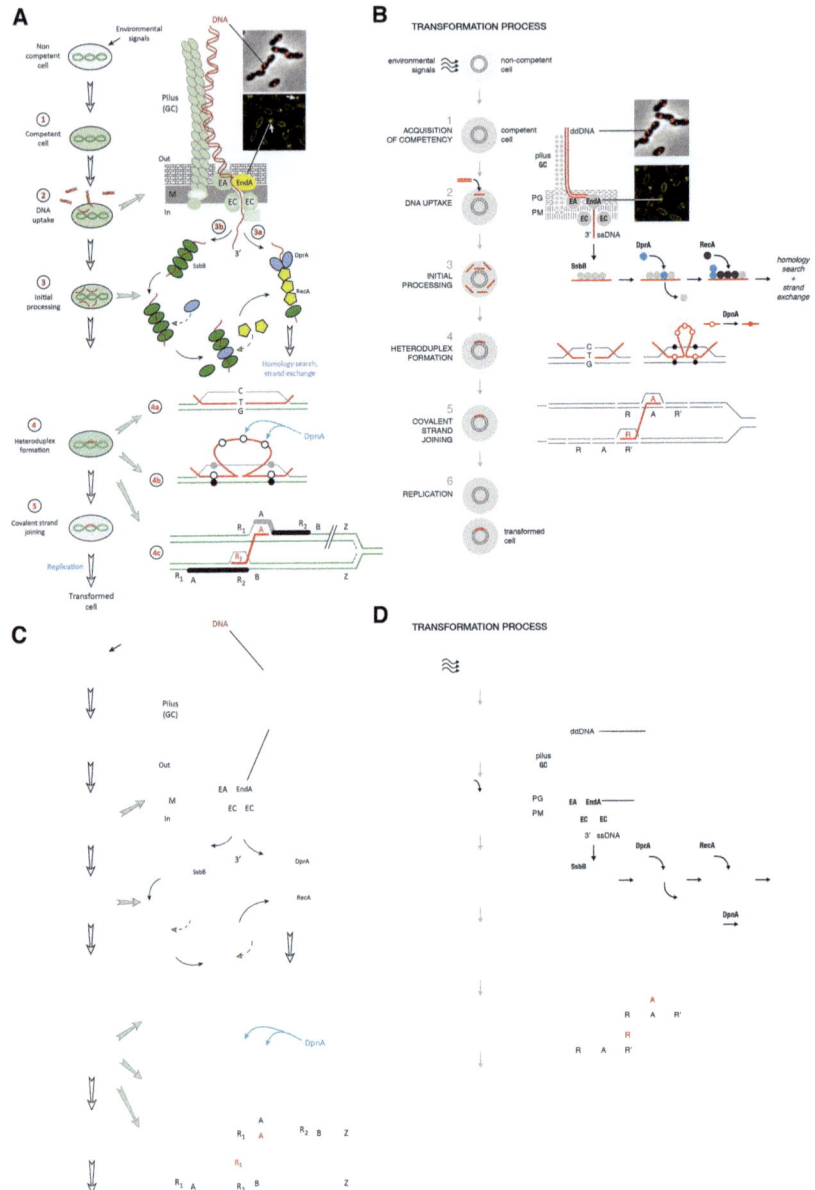

Plate 5 (Figure 3.4) Consistent alignment and judicious use of color and shape variation helps convey complex concepts and processes. **A.** Figures 1 and 2 from Johnston et al. (2014). **B.** Their redesign making use of alignment, simple shapes, and increased focus on the transforming dsDNA. **C.** Arrows, callouts, and entity labels from A. **D.** Arrows, callouts, and entity labels from B.

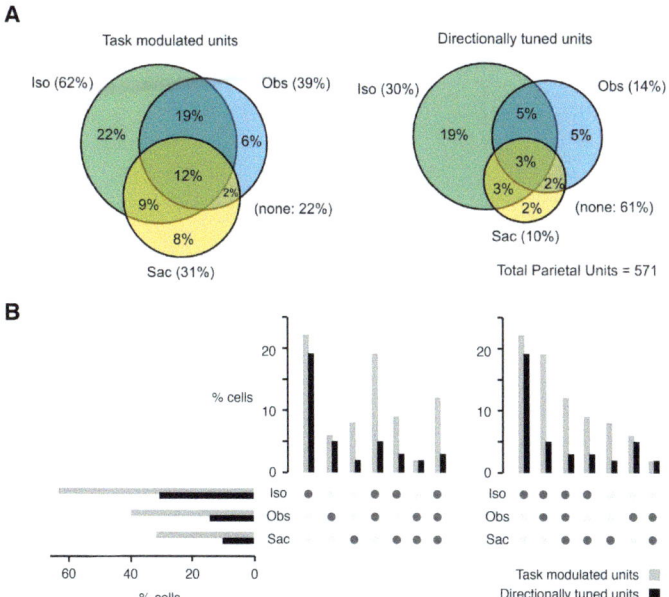

Plate 6 (Figure 3.5) Data encoding should be chosen to clearly demonstrate proportion. The use of areas to represent quantity should be avoided. **A.** Adapted from Figure 2 in Ferrari-Toniolo et al. (2015). Venn diagrams show proportion of cells in inferior parietal brain regions are significantly modulated (left) and directionally tuned (right) in three different tasks: isometric force production (Iso), visual observation (Obs), and saccadic eye movements (Sac). **B.** UpSet encoding of the data in Figure 3.5A.

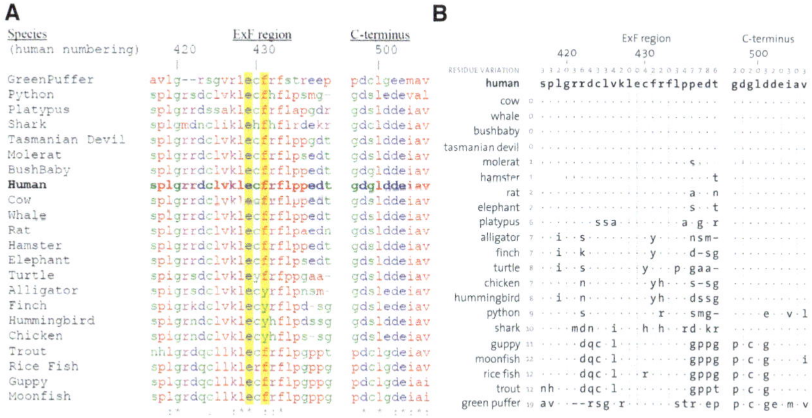

Plate 7 (Figure 3.6) Focus on differences helps identify patterns in data where small variation exists within a consensus. **A.** Adapted from Figure 8 in Li et al. (2015). The figure shows the evolutionary conservation of sequence across different species. **B.** Redesign of Figure 3.6A. Emphasis on differences is achieved by showing the human sequence first and in other species only those residues that are not conserved. The extent of differences is quantified by showing the number of different residues at a given position – in the original figure, this important quantity was encoded using symbols and relegated to the bottom of the figure. The order of species is based on the number of variants.

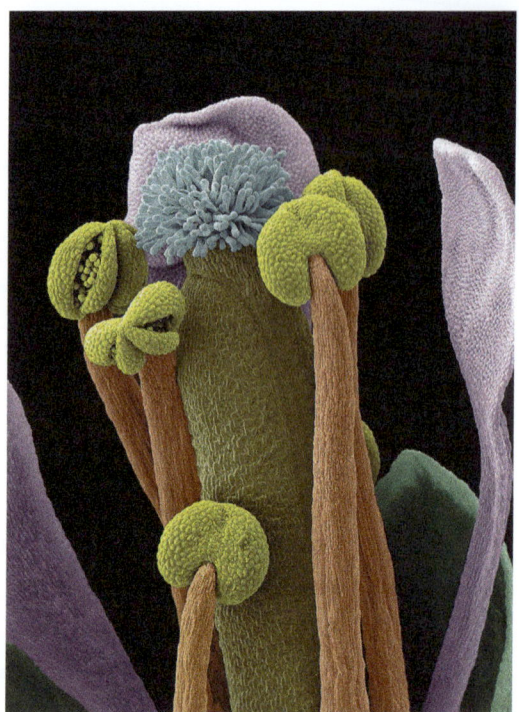

Plate 8 (Figure 4.1) Stefan Eberhard: *Plant reproductive parts*, Wellcome Images 2011. © Creative commons license. 'Scanning electron micrograph (SEM) of an Arabidopsis thaliana flower, also commonly known as thale cress. Some of the anthers are open, revealing pollen grains ready for dispersal. Arabidopsis was the first plant to have its entire genome sequenced and is widely used as a model organism in molecular and plant biology. Horizontal width of image is 1200 µm. Magnification 100x' (Wellcome, 2011).

Plate 9 Robert Mapplethorpe: *Poppy*, 1988. © The Robert Mapplethorpe Foundation Used by permission.

Plate 10 (Figure 5.1) The Orion nebula shown in a combination of infrared, visible, and ultra-violet light. The green hue stems from hydrogen and sulphur in the cloud, whereas the orange colors stem from complex organic molecules.

Image credit: NASA/JPL-Caltech/STScI.

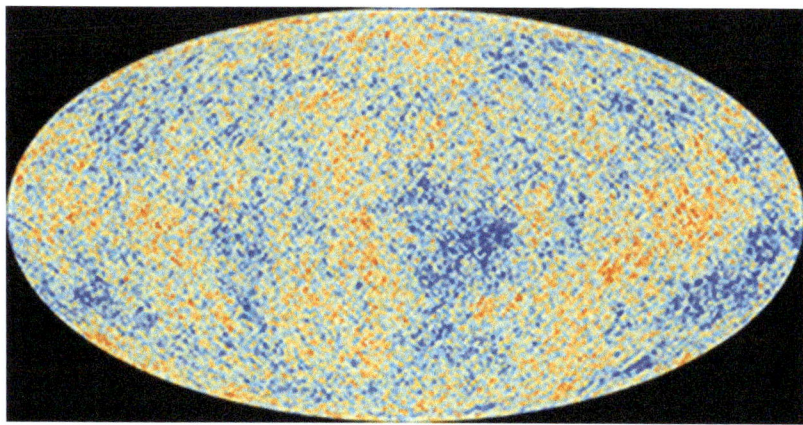

Plate 11 (Figure 5.3) The cosmic microwave background (CMB) as detected by the Planck satellite. The color-coding reflects the intensity of the radiation in various directions on the sky, with red signifying high intensity and blue corresponding to low intensity. The difference in intensity between the red and the blue regions is only about 1 part in 100,000.

Image credit: ESA and the Planck Collaboration.

Plate 12 (Figure 5.5) The Bullet Cluster measured in the visible spectrum and then printed as a collage with additional information measured in X-rays (pink) and through gravitational lensing (blue).

Image credit: NASA/CXC/CfA/Markevitch et al.

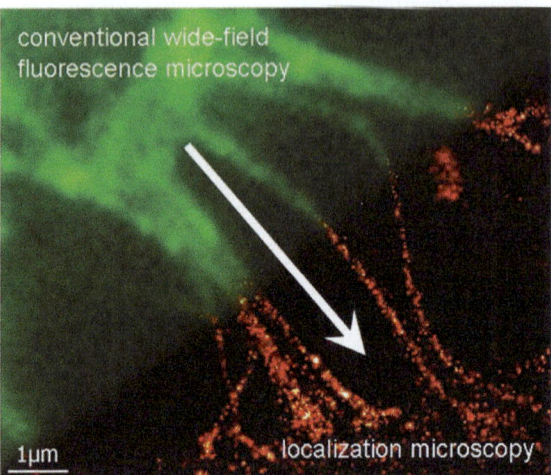

Plate 13 (Figure 6.1) Comparison between two microscopic visualization techniques: conventional wide-field fluorescence microscopy (upper left) and localization microscopy (lower right).

Source: From www.kip.uni-heidelberg.de/AG_Cremer/en/content/localization-microscopy-spdm. Image created by the C. Cremer research group of Applied Optics, Information Processing & Biophysics of Genome Structure, Heidelberg University based on data from P. Lemmer, M. Gunkel, D. Baddeley, R. Kaufmann, A. Urich, Y. Weiland, J. Reymann, P. Müller, M. Hausmann, and C. Cremer (2008) "SPDM: light microscopy with single-molecule resolution at the nanoscale". In *Applied Physics B, Lasers and Optics*. doi:10.1007/s00340-008-3152-x.

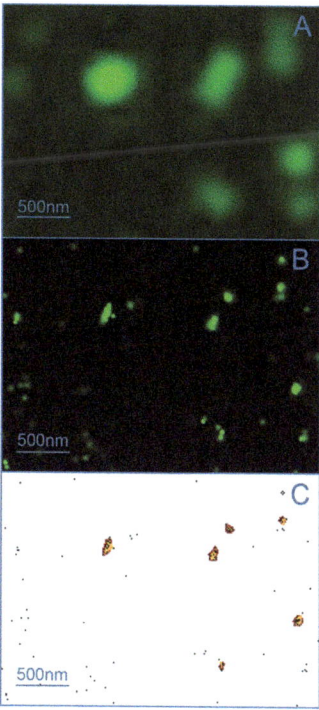

Plate 14 (Figure 6.5) Image comparison. Rainer Kaufmann et al. (2011, 50). **A.** Comparison between conventional wide-field fluorescent image, **B.** localization image of the same section, and **C.** results of the cluster finding algorithm.

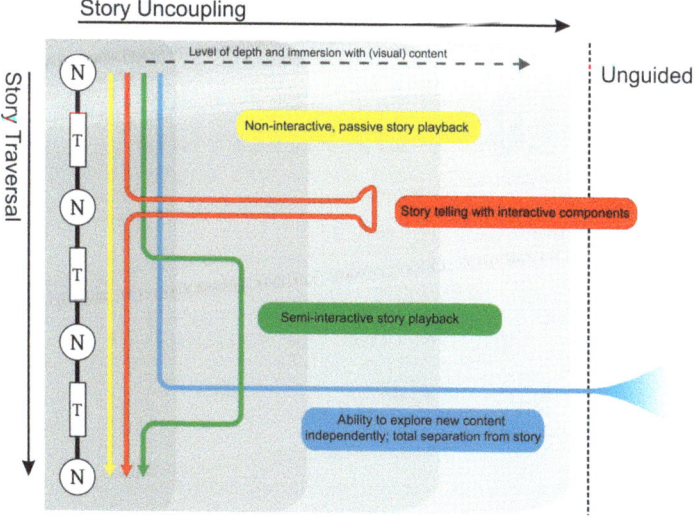

Plate 15 (Figure 8.3) Possible storytelling and interaction schemes, modified from Wohlfart and Hauser (2007) by Joep Veldhuis. Non-interactive, directed, and passive story consumption: Here, the reader is fully guided by a story plot from start to finish e.g. in a video (**yellow**). Storytelling with interaction: At a story node, the audience has the ability to halt the story, temporarily take control, and independently explore e.g. rotate fossil bones 360° on a screen (**red**). Semi-interactive playback: Users can take control, not just for a brief excursion, but for an entire section, skipping certain story elements in the process e.g. rather than finding about early human ancestors, skip straight to the Neanderthals (**green**). Total uncoupling from the story where users are allowed to alter the scene and engage in total freedom (**blue**). *N*: story nodes; *T*: story transitions.

Plate 16 (Figure 9.1) Anthropogenic electromagnetic noise disrupts magnetic compass orientation in a migratory bird. Storytelling with covers: The subject of the research paper featured on the cover is how anthropogenic electromagnetic noise disrupts the magnetic compass of migratory birds, like the European robin (pictured). We combined images of a specific sort of radio tower that creates the noise, electromagnetic waves, and a relevant bird species in flight to give the viewer a good idea of the story at a glance. © Macmillan Publishers Limited.

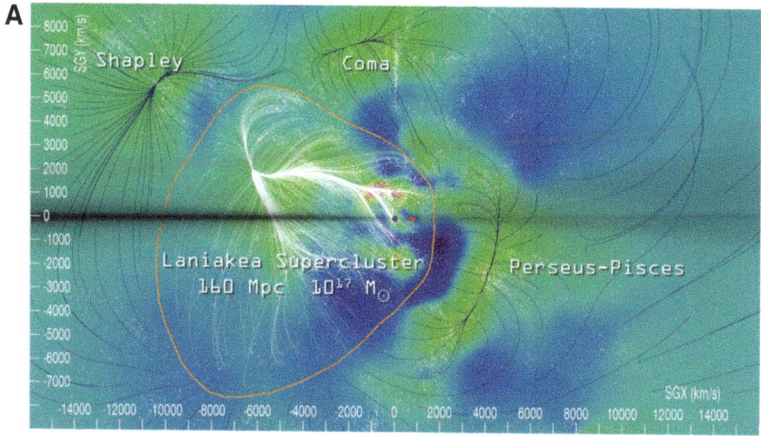

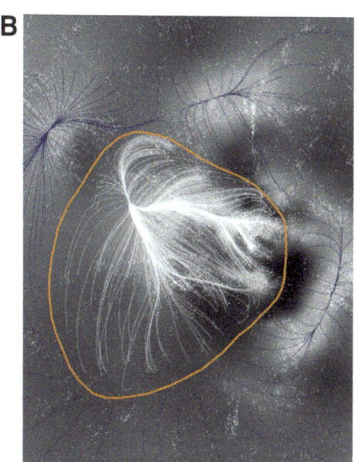

Plate 17 (Figure 9.2) The Laniakea supercluster of galaxies. The making of a cover: The author's original visualization of the supercluster (**A**); the author's second submission in response to our requests, to use as a reference for final artwork (**B**); and the final art created by a specialist astronomical illustrator (**C**). © Macmillan Publishers Limited.

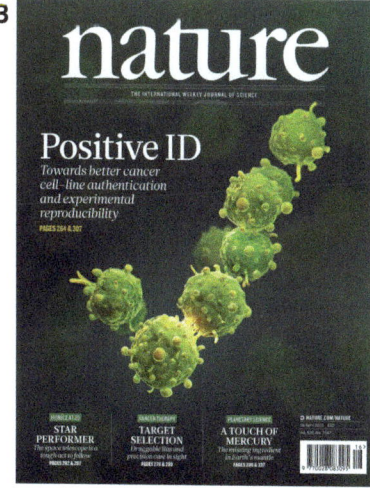

Plate 18 (Figure 9.3) A resource for cell line authentication, annotation, and quality control. Using visual metaphor: The author's submission (**A**); and the final artwork (**B**), a twist on the author's original question mark metaphor, rendered in 3D by a specialist medical illustrator. © Macmillan Publishers Limited.

 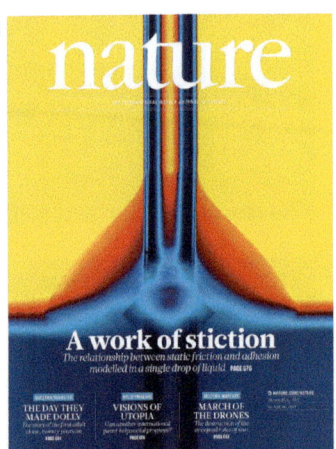

Plate 19 (Figure 9.4) A liquid drop on a solid surface. Scientific imaging: The author's original submission (**A**); and the final cover (**B**), created by the author in response to our request for an image of a single drop. The final image uses 390 sequential 1 MB photographs, with bold use of color that is appropriate in the context of a cover. © Macmillan Publishers Limited.

 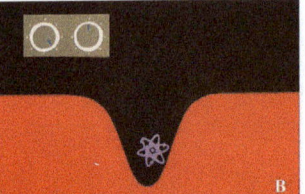 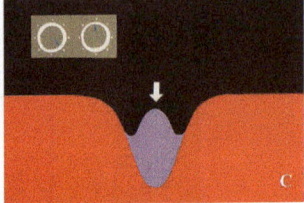

Plate 20 (Figure 10.1) Still images from the stop-motion animation *How your gameplay is used to help build a quantum computer* directed by Janet Rafner and Pinja Haikka. **A.** How quantum computers execute multiple calculations simultaneously. **B.** An atom in a movable energy well. **C.** The same atom as a wave function showing its most likely position. The wave function is a more scientifically accurate description of an atom.

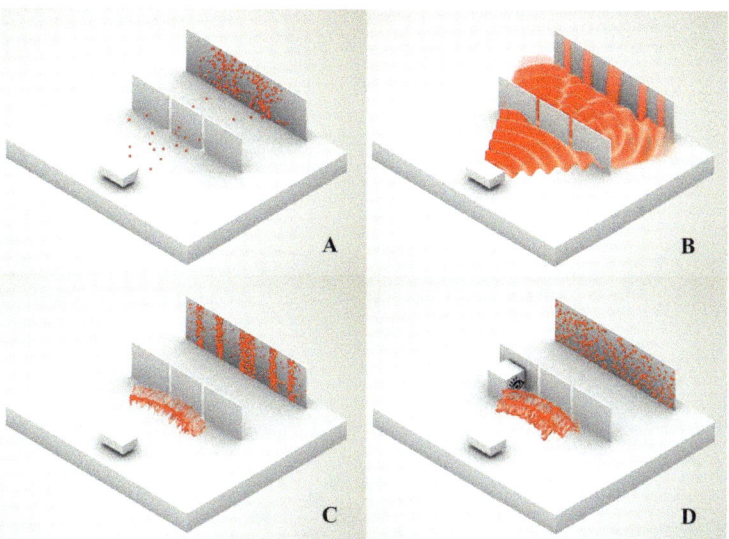

Plate 21 (Figure 10.2) *Quantum Made Simple*, traditional animation. Screenshots of the double slit experiment animation, displaying **A**. particles, **B**. waves, **C**. quantum wave functions, **D**. quantum wave functions with an observer at one of the two slits (resulting in the destructive interference pattern). Full animation available at www. QuantumMadeSimple.com.

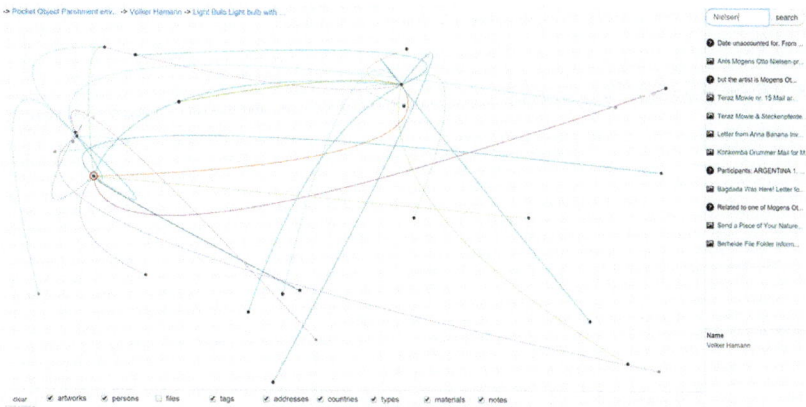

Plate 22 (Figure 11.2) The first prototype of a visual online database of the mail art collection made by Martin Luckmann (interactive designer) and Theis Vallø Madsen (researcher). Each search results in the same nodes, but the relational patterns are created randomly.

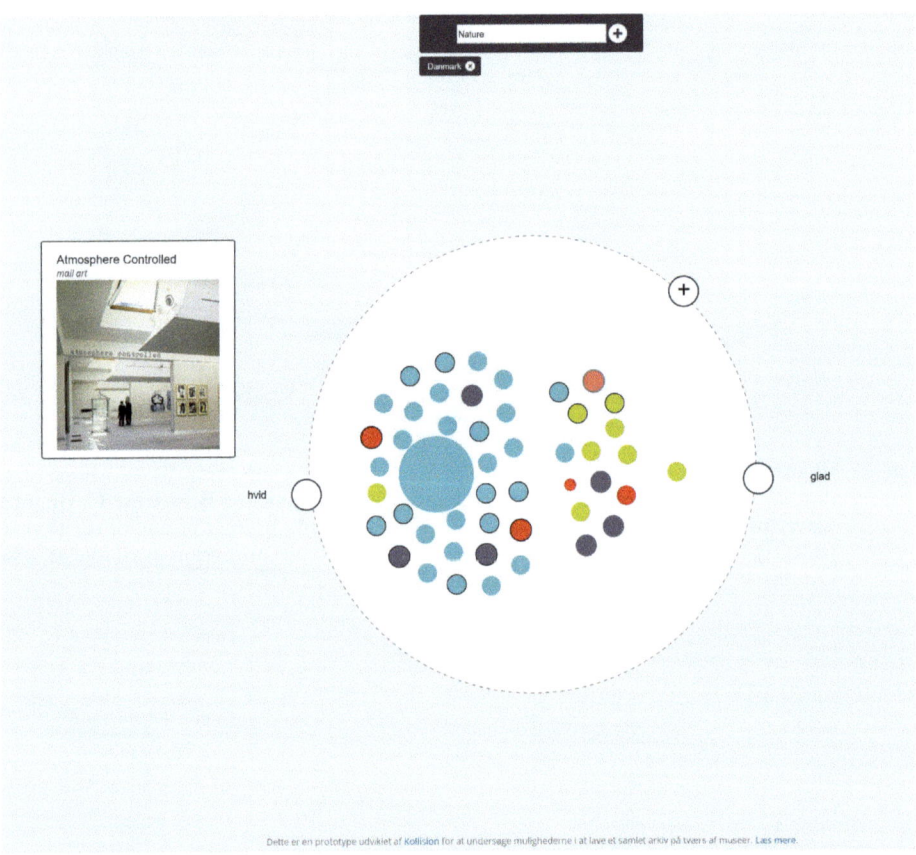

Plate 23 (Figure 11.3) The second prototype (Danish version) for a visual online database of the mail art collection. Items are attracted to certain categories. Colored dots (mail art pieces) are suspended between a magnet tool set on "hvid" (white) and a magnet tool set on "glad" (happy). Searching for "Nature" will pull out orange dots (specimens from the Museum of Natural History in Aarhus). Users can group items and add tags. Made by Kollision Aps and *The Mapping the Archive* project.

6   The Epistemics of Data Representation

How to Transform Data into Knowledge

Nina Samuel

Following the question of a visualization of the invisible of the previous chapter, this chapter zooms in from something that is very big and very far away (i.e. outer space) to something that is extremely small and as close to us as possible: microscopic images of human cells. Switching from the macro-cosmos to the micro-cosmos, we are confronted with a very similar problem: The phenomena are interpreted visually – given a concrete visual form – by accumulating and handling large amounts of mathematical data. In both cases, we are dealing with three distinct elements that are in a relation to each other but that need to be neatly distinguished: a phenomenon that is investigated (e.g. a star, a galaxy, a cell), data gained from this phenomenon through a technological device, and finally a representation – usually an image – based on an interpretation of the data (often referred to as "image processing" or "manipulation"). This chapter focuses on how these three elements relate to each other, and especially, how data measurements are transformed into visual representation and how they actively participate in the formation of theories. It is a key assumption of this chapter that representations have the power to actively influence the direction of scientific research and shape the development of scientific hypotheses and that representations can provide experts with new insights about their field (see Bredekamp, Dünkel & Schneider, 2015). This is called the epistemic function of representation. Images are no ornament in the scientific discourse; they are no retroactive visualization of existing facts. On the contrary, they are crucial for the decision on what becomes a "fact" and what is discarded from the scientific discourse.

The Quest for Increasing Resolution

This epistemic function of representation can be observed through centuries. The development of different microscopic techniques has informed our knowledge of cells' structures and functions at all times. Media employed to visualize these structures and technical expertise has shaped biological thought and our image of nature (cf. Bruhn, 2011). As a prominent example, the shift from electron microscopy to electronic light microscopy (i.e. video microscopy) has directed biologists' view of the cell away from the static view (where terms like "architecture" and "skeleton" were used for biological structures) to a new dynamic one and to a completely new understanding of living cells (Breidenmoser et al., 2010, 19–22). The following will ask in which way the recently invented method of localization microscopy contributes to an understanding of the epistemics of biological imaging and data visualization in general – and to the relation of the three elements mentioned above (phenomenon, data, and representation).

In a nutshell, the history of microscopy could be called an ongoing quest for increasing resolution. Localization microscopy is a super-resolution microscopic technique invented in 2006 that takes this process a step further by challenging the diffraction limit of traditional light microscopy (for the first decisive steps, see Betzig et al., 2006; Hess, Girirajan & Mason, 2006; Rust, Bates & Zhuang, 2006). Super-resolution microscopy was developed to make visible intracellular structures that are increasingly smaller than the limits of optical resolution, i.e. at this point in time structures that are below 10 nanometers small instead of 200 nanometers (Huang et al., 2016; Xu, Babcock & Zhuang, 2012; Diaspro, 2010). This limitation of light microscopy is commonly referred to as the Abbe-Reyleigh limit (Abbe, 1873). A common way to present and advertise the new microscopic technique to peers and to the interested public is to display a visual comparison between two images: A blurry region of a microscopic image representing the older technique and a sharply defined region of the same structure with fine granularity (Figure 6.1, Plate 13).

Here, an arrow points from an undifferentiated vagueness of darker and lighter green shadows toward an ornament of neatly composed little red dots against a black background. The arrow seems to indicate the direction of technological and epistemic progress, as symbolized in the transition from blurriness to visual acuity. The rhetoric used to promote this new visualization technique recalls a Christoph Columbian excitement:

> Once the light microscope opened up a door into a new world. But it was still impossible to reveal all the secrets of the microcosm: Structures smaller than 200 nanometers could not be made visible by light. But now, new methods allow us to transcend this frontier of spatial resolution that had previously been judged unconquerable and to break into the nanoworld of the cells.
>
> (Cremer, translation from German by author)

This rhetoric is typical for the introduction of a new visualization technique and, for example, had already been well known from the early days of computer visualization (Glaser, 1989). The increase of resolution is gained from the handling of a data matrix

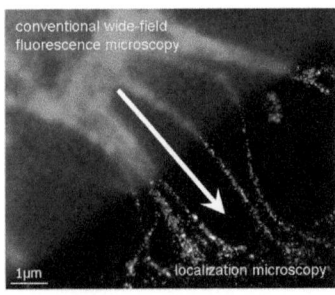

Figure 6.1 (See Color Plate 13) Comparison between two microscopic visualization techniques: conventional wide-field fluorescence microscopy (upper left) and localization microscopy (lower right).

Source: From www.kip.uni-heidelberg.de/AG_Cremer/en/content/localization-microscopy-spdm. Image created by the C. Cremer research group of Applied Optics, Information Processing & Biophysics of Genome Structure, Heidelberg University based on data from P. Lemmer, M. Gunkel, D. Baddeley, R. Kaufmann, A. Urich, Y. Weiland, J. Reymann, P. Müller, M. Hausmann, and C. Cremer (2008) "SPDM: light microscopy with single-molecule resolution at the nanoscale". In *Applied Physics B, Lasers and Optics*. doi:10.1007/s00340-008-3152-x.

consisting of numbers, yet it seems to allow us to "see better", to see in more detail, and hence, to also better understand the biological structure. New visualization techniques are often celebrated as leaps in understanding. As we have seen in the example of the astrophysical visualizations in Chapter 4 in this volume, scientific representations are trusted to help us to see something as it looks or even as it looked billions of years ago. But is this transition as smooth?

The Distance Between Phenomenon and Representation

A stunning fact described in the previous chapter is that astrophysical imagery – which is based on traveling light in various wavelengths – is forced to visualize the past. In other words: Astrophysics produces phantom images of phenomena that do not exist anymore. Because of its capacity to visualize the past, it might be worthwhile to discuss astrophysical imagery with reference to the notion of trace. A trace is a hint for something that is already gone, that is not yet visible anymore. In image theory, traces have most often been ascribed to photography (Geimer, 2007). Traditionally, photography is known as a medium suitable for preserving the memory of phenomena that have ceased to exist, but is the situation different if things are already nonexistent in the moment they are visualized? A trace can be called an antecedent manifestation of something that is not yet definable as image, text, or another form of representation (Rheinberger, 2007). Yet – and different from astrophysical images – a trace is often characterized by a material participation with the thing that left the trace. It is questionable whether this can be said for planets or galaxies that do not exist anymore when their emitted light is recorded.

This ambiguity points to the basic question of the intricate relationship between the representation and the actual phenomena that is represented. A fundamental distance characterizes this relationship. To introduce a rough distinction, this distance is often distinguished based on its indexical, iconic or symbolic quality (loosely based on Peirce, 1983). Obviously, this differentiation suggests the existence of ideal types (or archetypes) while the analysis of real imagery leads to a multitude of hybrid types. If the relationship between the object and the representation is based on physical presence or direct touch, it can be referred to as an indexical relationship (e.g. analog photographs or actual material casts). Although touch is not the only characteristic of these images, the notions of imprint and contact are strongly associated with them (Didi-Huberman, 1999; Dünkel, 2010). This implies a direct correlation in a given context and/or a kind of sensory feature that points to the thing that it references. In a symbolic relationship, on the contrary, the context is removable, and the meaning is determined through conventions (Peirce, 1983, 65). Diagrams (cf. Chapter 2, this volume) are often discussed as examples for symbols.

In both the case of astrophysical visualizations like cosmic microwave background (Chapter 5, this volume) and localization microscopy, the relationship between the phenomenon and the representation is neither fully symbolic nor fully indexical, because the data hold the intermediary role between the two. The meaning of the term "representation" – in sharp contrast to the reflection of some kind of "reality" – has to be scrutinized against this background (Levine, 1993; Hagner, 1997; Lynch & Woolgar, 1990; Rheinberger, 2001).

In the transformation process from phenomenon to representation, both the phenomenon-data-relation and the data-representation-relation exhibit contingencies. For example, the collection of data is influenced by the choice of a specific treatment of the specimen under investigation, on the laboratory setting, on the algorithms, models, technical devices used in the data acquisition process, and informed by theories that are all determined by their own historicity. Once the data are assembled, they mediate between phenomenon and representation and partake in both, yet one particular set of data can still result in numerous different representations. For example, the assignment of colors that is necessary to fabricate an image from the data is often arbitrary yet underpinned by consensus, as is the selection of detail or potential shading. Hence, there is no mandatory or forced causal relation between phenomenon and representation. Nevertheless, we assume that there is a direct and mimetic relationship between the galaxy or the cell and the image of the galaxy or the cell generated from the data. One could call this assumed relationship iconic, a term that embraces all kinds of different techniques of contemporary visualization practices that *make* something visible in a way so that it seems to mimetically resemble the phenomenon. Yet different from the understanding that an iconic relationship implies a pattern that physically resembles 'what it stands for' (Peirce, 1983, 64), it is important to emphasize the word "make" in this context: Data representation has to be understood as an *active* production, never as a passive *re*production.

This basic insight coincides with the definition of scientific representation as characterized by historian of science Hans-Jörg Rheinberger 'in terms of a production in which the depicted object takes shape in the first place' (Rheinberger, 1992, 73). Data representation belongs to those "experimental systems" that Rheinberger has investigated intensively. He notes that these systems display 'spaces of representation' where 'graphemes, material traces, are produced, are articulated and disconnected, and are placed, displaced, and replaced. Science "thinks" within its spaces of representation, within the hybrid context of the available experimental system' (Rheinberger, 1998, 287, 297). This is another description of the epistemic function of representations. Even if data are usually considered to exist in an immaterial form (while it is dubitable that this is true without restrictions; see, for example, Trogemann, 2010), the same characteristics of all experimental systems apply to them: Instead of displaying an "objective reality" of the thing or phenomena in question, they create 'things that otherwise cannot be grasped as scientific objects' (Rheinberger, 1998, 297). This applies both to visualization practices that *aim at analyzing* a phenomenon in the present moment of its existence (including cells in microscopic magnification or astrophysical phenomena that have already ceased to exist but whose analyzed data are still referring to its existence) and to those that *aim at predicting* a phenomenon. The latter are based exclusively on mathematical models (Hinterwaldner, 2010, 25) and can be called simulations. However, this "taking shape of the represented object" can better be understood if we step *beyond* images like Figure 6.1 that are produced to communicate and popularize the benefits of the new imaging technique.

Images of Image-Making

To obtain an image in localization microscopy, the visualization process is split in two parts: data acquisition and a subsequent image generation (Figure 6.2A and B). Both are temporal procedures. In the very beginning, energy (i.e. light) has to be added to the cell to provoke a fluorescent reaction. In order to do so, the biological material has to be activated with a laser beam. During this first phase, the molecules are literally "switched on and off", oscillating in intervals between fluorescent and reversible "dark" states. These temporary non-fluorescent states of single molecules

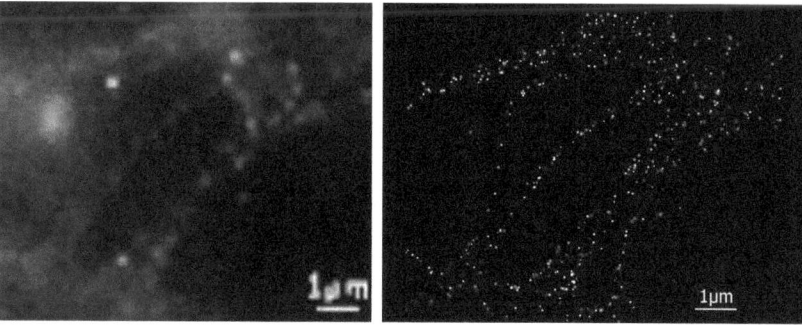

Figure 6.2 **A.** *Data acquisition* in localization microscopy. **B.** *Image processing* in localization microscopy. Rainer Kaufmann, filmstill (2009).

lead to their optical isolation and enable the acquisition of position data precisely down to a few nanometers (Kaufmann et al., 2010, 348).

Figure 6.2A shows a still from this procedure in the form of a silent black-and-white movie of a shadowy and blurred structure filled with blinking white dots. While the background structure originates from the texture of the cell, blinking points represent the stochastically activated molecules. The image stems from the first phase of the image-making process, exclusively used during research and in order to document and analyze data acquisition. The interim result of this procedure is a list of discrete coordinates containing information about estimated positions, their statistical accuracy, their quantity, and the shape of the detected spots. Taken together, this information constitutes a matrix of data that can grow up to the gigabyte range.

Similar to the image processing of cosmic microwave background (Chapter 5, this volume), the processing of this enormous volume of data poses a major computational challenge. Over a longer period of computing time, the recorded position data are assembled into a new unified graphical space (Figure 6.2B). However, this structure that now resembles the typical style of the advertised "sharp" grainy structure of localization microscopy (cf. Figure 6.1, right side) is not just there all at once. On the contrary, slowly and point after point, the array of numbers has to be translated into spatial relations and colors. Statistical methods (usually a Gaussian rendering) have to be applied to transform the measured positions of the blinking dots into an image displaying the information sought in biological research.

Images from the image-making process and a close reading of the production process in localization microscopy teach us three important features of data representation that help to give answers to what mechanisms play a role to transform data into knowledge. They can be summarized with three keywords that will be exemplified in the following: composition, emergence, and continuity.

Composition

Unlike older microscopic techniques, like conventional wide-field fluorescence microscopy, the "final" localization microscopy image used to communicate the scientific results (Figure 6.1, right side) cannot be viewed directly by the scientist's eye and captured by a single exposure. On the contrary, this image is a composite, a data collage (cf. Buschhaus & Hinterwaldner, 2006). The only images that can actually be observed during the scientific process are the images of the image-making process: the measuring (Figure 6.2A) and the assembling (Figure 6.2B). The composition is put together during this temporal procedure, dot after dot (Figure 6.2B). Eventually, every single measurement appears in the same homogeneous pictorial space, disguising its composite nature.

Different from the paradigm of observation that ruled the natural sciences in the 19th century (Hoffmann, 2006), we are witnessing a shift: Now observation is replaced by a measuring device and a data acquisition process, but seeing is still key when it comes to the interpretation of the massive amounts of data. Both in astrophysics and localization microscopy, the scientist relies on the eye to detect (ir-)regularities, structures, and patterns in the data that are interpreted visually. This practice is a variation of what historians of science Lorraine Daston and Peter Galison described as "trained judgement" in their study on objectivity (Daston & Galison, 2007). Different from what they call "mechanical objectivity", which is characterized through an "automated transfer", localization images are the result of a lengthy data processing requiring many individual choices and alterations. The "expert-scientists" are aware of the 'intrusion of the subjective into the images' (Daston & Galison, 2007, 370) when they perform direct interventions to obtain a representation (e.g. the activation of the cell with the laser beam). Because of the composite nature of localization images, researchers must have to learn to read the images and use their "trained judgment" to be able to explore them as source of information about the cell. Even though observation has shifted away from the phenomenon to the data, like all epistemic practices, the role of the eye is not diminished. The eye does not investigate the object directly, nor does it access the object through an optical device that produces an enlarged representation, yet it scrutinizes the data that are made accessible in the form of a visual representation – or optionally the ear if the representation is sonic (see Chapter 7, this volume).

Emergence of the Unexpected

The production of a localization image is no genuinely natural procedure and can have unpredictable and surprising effects on what is ultimately depicted – a new phenomenon of undeterminable origin can emerge accidentally in the images. These phenomena are often called artifacts. Overall, there are three aspects in the transition from cell sample to representation that harbor the unexpected appearance of artifacts: first, the chemical cell preparation; second, the activation of the fluorescent molecules in the cell with the laser beam, i.e. the recording process; and third, the subsequent algorithm used to calculate the exact positions of the molecules and depict them in the final image.

First of all, even though the goal of most microscopic methods might be the mere visualization and analysis of a given structure without changing the structure itself, in a way one is inevitably forced 'to intervene if one decides to represent' (Hacking, 1983). Since the earliest days of microscopy, preparation of each and every specimen necessarily modified the biological material under observation to a certain extent. However, localization microscopy is a drastic example for this insight. In order to obtain an image with the highest spatial resolution in localization microscopy, the cell has to be chemically fixed. A living cell contains too many fast-moving elements to perform the necessary localization measurements over time. However, the process of chemical fixation is associated with structural changes in the sample (Bleck et al., 2010; Schnell et al., 2012; Kaufmann et al., 2014; Johnson et al., 2015).

Second, how images are obtained differ radically from older methods like classical bright-field microscopy: To be able to collect the necessary data, a real-life activation of photo-physical reactions inside the biological material is needed. In order to be ready for "observation", the biological material has to actively respond to the treatment. The laser beam causes a phyto-physical reaction inside the cell, and activating fluorescence through the data acquisition process alters the electronic states and configurations of the molecules. Hence, unintentional interactions between the imaging process and the molecules are a genuine part of the research. In the reality of the laboratory, this can have startling effects: Sometimes, 'very bright spherical objects' (Kaufmann et al., 2010, 349; see also Samuel, 2013) that cannot be understood with the ruling theories suddenly appear during a series of experiments and reveal the ambivalence between scientific investigation and artifact. The emergence of such unexpected visual "objects" is an integral part of the visualization process, as it can easily happen that the interaction of the laser beam (i.e. the photons) with the molecules produces additional blinking spots, or that molecules, which were not originally addressed, react.

Figure 6.3 is an example of these kinds of unexpected emergences. The one scattered white spot in the middle of the egg-shaped dark zone (i.e. the cell nucleus) should actually not be there according to the theoretical background of the study. To understand what its origin could be, it is important to understand another detail of the production process of localization microscopy. To provoke the fluorescent reaction necessary to produce the images, there are two common ways: Either DNA has to be inserted that makes the cell "fabricate" the target protein with a connected fluorescent protein or antibodies with fluorescent molecules have to be inserted who will then connect to the target proteins. Now, if white spots appear at locations where they were not expected, this can have two reasons: First, one of the fluorescent molecules might have landed at some place where it was not intended and thus could have

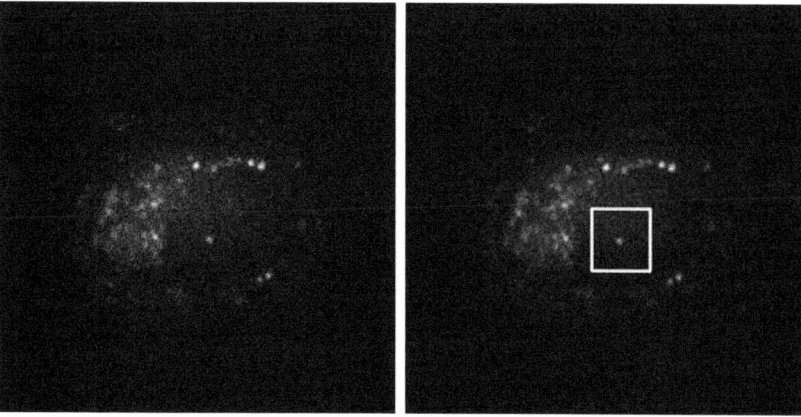

Figure 6.3 Images from the raw data stack of the localization microscopy procedure from the research of Rainer Kaufmann (2016). The same image as Figure 6.2A but with light spot marked. Rainer Kaufmann, film-still (2009).

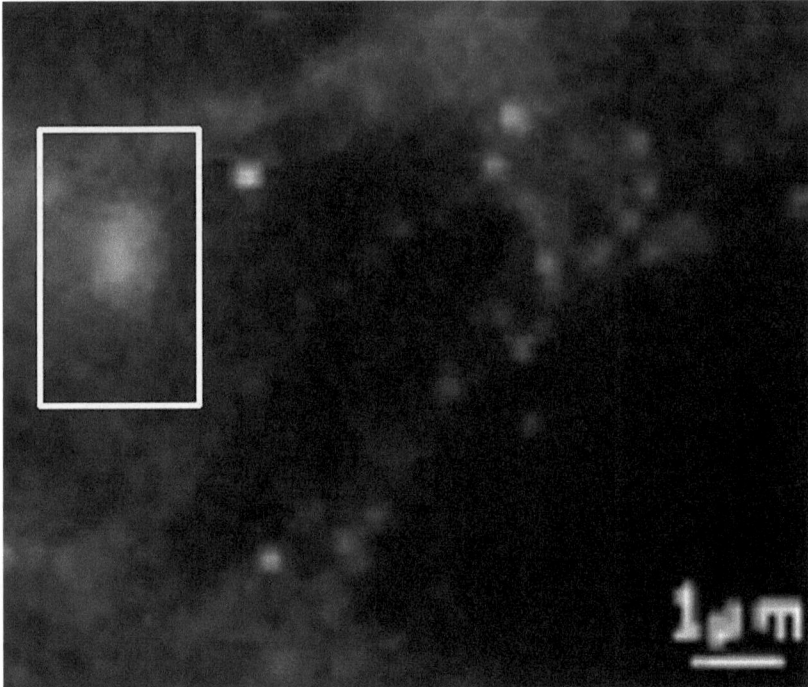

Figure 6.4 Data acquisition in localization microscopy.

provided an unspecific signal. Or second, an unmarked molecule might have shown, for reasons unknown, a blinking behavior similar to the marked molecules. While it is possible to perform subsequent control experiments to minimize the emergence of these inexplicable dots, it is very hard to erase all of them.

However, besides those unexpected emergence of blinking molecules, some spots in localization images can also have different origins. The larger, slightly blurred white spot in the left side of the image of Figure 6.4 is such a case. While in the original movie where this still is taken from, the surrounding spots are flashing lights, constantly blinking and full of bustling activity, this one spot remains unchanged during the whole process. There are two possible explanations for this spot: First, it could be an actual trace of the instrumental setting and the recording process that mediates between the cell and the representation, a visual reminder that the production process has a direct impact on the morphology of the image. The appearance of the instrumental setting is another type of unintended emergence that is usually excluded or retrospectively erased in images that are used to communicate science.

The second explanation for this spot is related to the recording process as well: It could be an out-of-focus signal. One of the unwanted side effects of depicting structures that are located deeply inside the cell is the existence of fluorescent light outside of the focus area. It is a similar effect that you can observe on photographs taken with a large aperture and long focal distance – fuzzy white spots. If the algorithm does not filter them out, it can lead to problems localizing the proper signals.

All these aspects taken together, the scientists are confronted with the question whether they see something that contains new information about the actual cellular structure or just a random and undesired effect of the interaction between the cell, the chemicals, the recording device, and the algorithm. Images from the image-making process display this difficulty: They are typically images of doubt.

On a representational level, these inevitable peculiarities of the microscopic imaging process constitute zones of insecurity that distinguish epistemic processes in experimental systems. When Rheinberger defines the "objects of inquiry" of such experimental systems, he notes that they 'present themselves in a characteristic, irreducible vagueness' (Rheinberger, 1997, 28). This vagueness or insecurity is also characteristic for the phenomena that are investigated with the method of localization microscopy and must be understood as a productive force in research. It opens the realm of the unexpected and enables the sudden emergence of phenomena triggering confusion, revision, or new insights.

Continuity

While the eye is often the preferred organ used to scrutinize data, the specific format made accessible for the eye is not fixed. Data are aniconic, and hence they always have to be interpreted and displayed to be accessible to the senses. Therefore, the researcher must make a conscious choice, when visualizing data, determined by the communicative function that data representation has to fulfill in the scientific community. Scientific representations have to be able to migrate through different disciplinary audiences – in the case of localization microscopy between scientists who produce the images and scientists who use them for their own research. During this process of interdisciplinary collaboration, and especially valid in the first years after the introduction of localization microscopy, expectations of how the investigated biological structure is supposed to look are exchanged and adjusted. Figure 6.5 (Plate 14) compares three different types of visualization of the same biological section.

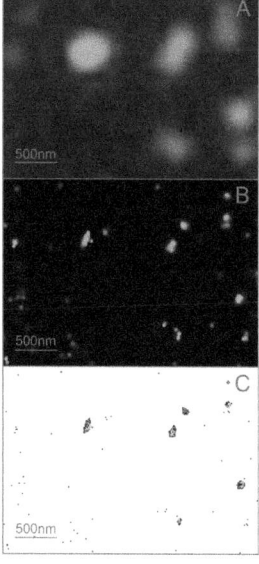

Figure 6.5 (See Color Plate 14) Image comparison. Rainer Kaufmann et al. (2011, 50). **A.** Comparison between conventional wide-field fluorescent image, **B.** localization image of the same section, and **C.** results of the cluster finding algorithm.

Figure 6.5 (Plate 14) shows the conventional wide-field fluorescent image that resembles hazy bright blotches embedded in a black setting (A) and, at the bottom, a localization image showing the result of the cluster-finding algorithm, composed of sharp dark spots against a white background (C). According to the physicist Rainer Kaufmann who produced the image, this last image comes closest to an "accurate" visual representation of the measured data (Kaufman, 2011, interview). However, he stated, if the data were handed over to the biologists in this specific pictorial form, the image would probably cause an interdisciplinary communication blackout: Unable to read it, the biologists might just as well return the images to the physicist. In other words, and with reference to Daston and Galison's notion of "trained judgment", the training process that scientists need to undergo before they are able to actually decode and judge new imagery happens at a slower pace than the speed with which new technology is made available for research. Therefore, based on the striking visual discrepancy between Figure 6.5A and C, and in order to become comprehensible for the biologists, localization images are reworked digitally in order to imitate the older wide-field view as closely as possible, as can be seen in the middle image (Figure 6.5B). In some cases, the sharpness of the data measurement even has to be retroactively "smudged" to familiarize the biologist with the unknown – although the physicist utterly regrets the loss of information density. Conventions of seeing, the habituation of the eye, visual expectations and continuity between older and newer visualization techniques are key when it comes to the decision in which way the data measurements take shape as an image. Aesthetics and epistemics have to collaborate to circulate scientific results between different scientific cultures.

Scientific Data Representation as Thought Instrument

Scientific images are instruments for thinking. As such, they have to be understood as both products and producers of knowledge. Composition, emergence, and continuity are only three of many traits of representation that play a role when data are to be transformed into new insights, but they exemplify the intricate relationship between phenomenon, data, and representation that produces innumerable hybrid forms between the indexical, symbolic, and iconic imagery.

A composition does not emerge all at once but step by step over time. It is only the *process* of representation that makes the object under investigation *accessible*. Moreover, and most important, it is these images from within the research process (Figures 6.2A, 6.3, 6.4) – from the workbench and hidden from a larger audience – that reveal the epistemic force that images can have for the scientific thought (cf. Nasim, 2013; Samuel, 2012; Holmes, Renn & Rheinberger 2003). Key for the understanding of the epistemics of data visualization is a deeper understanding of this procedural nature. It is the complexity of this process that fosters the emergence of unexpected details in data representations. It is the embedding of this process in the culture of experimental systems (Rheinberger, 1998) that fosters the necessity of migrating images. It is the stylistic continuity of scientific representations that facilitates this migration between different scientific audiences.

Different practices result not only in different cultures of seeing but also in different interpretations of the reconstructed microscopic vision. However, pictorial norms and conventions function as a solid filter in two distinct directions: First, they influence the way data are translated into an image at all; second, they determine the frontier between the expected and the unexpected, the accepted and the contested result. Our eyes have the irresistible tendency to be inclined to see and recognize images that they are familiar with (Bachelard, 1974, 172). This has the effect that an unexpected appearance of an unclassifiable shape might be discarded because it contradicts trained ways of seeing even though it contains a new insight. New image processing techniques can open new spaces of knowledge in the same way as they can limit understanding.

The history of scientific images is also a history of the construction of the visible from the invisible (see, for example, Chapter 5, this volume). The constant inquiry of questions like, "When does something appear, when does it disappear, and when does one decide that it is 'visible enough' to be called a new phenomenon – or even a 'discovery' or 'fact'?" affects the relationship between image processing technique, image, and knowledge. Images that display these thresholds of visibility and reveal the unavoidable moments of insecurity and doubt circulate within the scientific community during the phase of research and production but are often discarded as soon as the desired "final result" is obtained (e.g. Figure 6.1, right side). A strong focus on the "final result" implicitly disregards the original questions that drove the research, but if science wished to be critical, it may benefit from disclosing and analyzing the images from within its research process because they reveal the epistemic qualities of images in science in the strongest sense. Images in science are not only illustrations of work done by other means to communicate scientific results; they can also be regarded as tools for experimental discovery in their own right.

References

Abbe, E. 1873. "Beiträge zur Theorie des Mikroskops und der mikroskopischen Wahrnehmung". *Archiv für Mikroskopische Anatomic*, 9: pp. 413–418.

Bachelard, G. 1974. *Epistemologie. Ausgewählte Texte.* Frankfurt am Main, Berlin, Vienna: Ullstein.

Betzig, E., G. H. Patterson, R. Sougrat, O. W. Lindwasser, S. Olenych, J. S. Bonifacino, and M. W. Davidson. 2006. "Imaging Intracellular Fluorescent Proteins at Nanometer Resolution". *Science* 313(5793): pp. 1642–1645.

Bleck, C. K., A. Merz, M. G. Gutierrez, P. Walther, J. Dubochet, B. Zuber, and G. Griffiths. 2010. "Comparison of different methods for thin section EM analysis of *Mycobacterium smegmatis.*" *Journal of Microscopy* 237: pp. 23–38.

Bredekamp, H., V. Dünkel, and B. Schneider (Eds.). 2015. *The Technical Image: A History of Styles in Scientific Imagery.* Chicago: University of Chicago Press.

Breidenmoser, T., F. O. Engler, G. Jirikowski, M. Pohl, and D. G. Weiss. 2010. "Transformation of Scientific Knowledge in Biology: Changes in our Understanding of the Living Cell through Microscopic Imaging". *Max Planck Institute for the History of Science: preprint 408.*

Bruhn, M. 2011. "Life lines: An art history of biological research around 1800". In *Studies in History and Philosophy of Science Part C: Studies in History and Philosophy of Biological and Biomedical Sciences* 42(4): pp. 368–380.

Buschhaus, M. and I. Hinterwaldner (Eds.). 2006. *The Picture's Image. Wissenschaftliche Visualisierung als Komposit.* Munich: Wilhelm Fink Verlag.

Cremer, C. "Vorstoß in den Nanokosmos–Neue Mikroskope überschreiten für unüberwindlich gehaltene Grenzen". Cited 2016 March 28. www.uni-heidelberg.de/presse/ruca/ruca08-3/vorst.html.

Daston, L. and P. Galison. 2007. *Objectivity.* New York: Zone Books.

Diaspro, A. 2010. *Optical Fluorescence Microscopy from the Spectral to the Nano Dimension.* New York: Springer.

Didi-Huberman, G. 1999. *Ähnlichkeit und Berührung. Archäologie, Anachronismus und Modernität des Abdrucks.* Köln: DuMont.

Dünkel, V. (Ed.). 2010. "Kontaktbilder". In *Bildwelten des Wissens* 8,1. Berlin: Akad.-Verl.

Geimer, P. 2007. "Image as Trace: Speculations about an Undead Paradigm". *Differences* 18(1): pp. 7–28.

Glaser, P. 1989. "Das Kolumbus-Gefühl. Entdeckungen in einer virtuellen Welt". In *Computerkultur, oder, The Beauty of Bit & Byte,* edited by M. Weisser, pp. 19–40. Bremen: TMS.

Hacking, I. 1983. *Representing and Intervening.* Cambridge: Cambridge University Press.

Hagner, M. 1997. "Zwei Anmerkungen zur Repräsentation in der Wissenschaftsgeschichte". In *Räume des Wissens. Repräsentation, Codierung, Spur,* edited by H. J. Rheinberger, M. Hagner, and B. Wahrig-Schmidt, pp. 339–355. Berlin: Akademie.

Hess, S. T., T. P. K. Girirajan, and M. D. Mason. 2006. "Ultra-High Resolution Imaging by Fluorescence Photoactivation Localization Microscopy." *Biophysical Journal* 91(11): pp. 4258–4272.

Hinterwaldner, I. 2010. *Das systemische Bild: Ikonizität im Rahmen computerbasierter Echtzeitsimulationen.* Munich: Wilhelm Fink Verlag.

Hoffmann, C. 2006. *Unter Beobachtung: Naturforschung in der Zeit der Sinnesapparate.* Göttingen: Wallstein.

Holmes, F. L., J. Renn, and H. J. Rheinberger (Eds.). 2003. *Reworking the Bench: Research Notebooks in the History of Science,* 1. Dordrecht: Kluwer Academic.

Huang, F. G. Sirinakis, E. S. Allgeyer, L. K. Schroeder, W. C. Dium, E. B. Kromann, T. Phan, F. E. Rivera-Molina, J. R. Myers, I. Irnov, M. Lessard, Y. Zhang, M. A. Handel, C. Jacobs-Wagner, C. P. Lusk, J. E. Rothman, D. Toomre, M. J. Booth, and J. Bewersdorf. 2016. "Ultra-high resolution 3D imaging of whole cells". *Cell* 166(4): pp. 1–13.

Johnson, E., E. Seiradrake, E. Y. Jones, I. Davis, K. Grünewald, and R. Kaufmann. 2015. "Correlative in-resin super-resolution and electron microscopy using standard fluorescent proteins". *Scientific Reports* 5, 9583: pp. 1–8.

Kaufmann, R., P. Müller, M. Hausmann, and C. Cremer. 2010. "Imaging label-free intracellular structures by localisation microscopy". *Micron (Oxford, England)* 42(4): pp. 348–352.

Kaufmann, R., P. Müller, G. Hildenbrand, M. Hausmann, and C. Cremer. 2011. "Analysis of Her2/neu membrane protein clusters in different types of breast cancer cells using localization microscopy". *Journal of Microscopy* 242(1): pp. 46–54.

Kaufmann, R. 2011. Interview led by Nina Samuel, Heidelberg, February 16, 2011 (unpublished).

Kaufmann, R., P. Schellenberger, E. Seiradake, I. M. Dobbie, E. Y. Jones, I. Davis, C. Hagen, and K. Grünewald. 2014. "Super-resolution microscopy using standard fluorescent proteins in intact cells under cryo-conditions". *NanoLetters* 14: pp. 4171–4175.

Levine, G. (Ed.). 1993. *Realism and Representation.* Madison: University of Wisconsin Press.

Lynch, M. and S. Woolgar. 1990. *Representation in Scientific Practice.* Cambridge: MIT Press.

Nasim, O. W. 2013. *Observing by Hand. Sketching the Nebulae in the Nineteenth Century.* Chicago: The University of Chicago Press.

Peirce, C. S. 1983. *Phänomen und Logik der Zeichen.* Frankfurt a.M.: Suhrkamp.

Rheinberger, H. J. 1992. *Experiment, Differenz, Schrift. Zur Geschichte epistemischer Dinge.* Marburg/Lahn: Basilisken-Presse.

Rheinberger, H. J. 1997. *Toward a History of Epistemic Things: Synthesizing Proteins in the Test Tube.* Stanford: Stanford University Press.

Rheinberger, H. J. 1998. "Experimental Systems, Graphematic Spaces". In *Inscribing Science. Scientific Texts and the Materiality of Communication,* edited by T. Lenoir, pp. 285–303. Stanford: Stanford University Press.

Rheinberger, H. J. 2001. "Objekt und Repräsentation". In *Mit dem Auge denken: Strategien der Sichtbarmachung in wissenschaftlichen und virtuellen Welten,* edited by B. Heintz, A. Benz, and J. Huber, pp. 55–61. Zürich: Edition Voldemeer.

Rheinberger, H. J. 2007. "Spurenlesen im Experimentalsystem." In *Spur. Spurenlesen als Orientierungstechnik und Wissenskunst*, edited by S. Krämer, W. Kogge, and G. Grube, pp. 293–308. Frankfurt am Main: Suhrkampf.

Rust, M. J., M. Bates, and X. Zhuang. 2006. "Sub-diffraction-limit imaging by stochastic optical reconstruction microscopy (STORM)" *Nature Methods* 3(10): pp. 793–796.

Samuel, N. (Ed.). 2012. *The Islands of Benoit Mandelbrot: Fractals, Chaos, and the Materiality of Thinking*. New York: Yale University Press.

Samuel, N. 2013. "Images as Tools. On Visual Epistemic Practices in the Biological Sciences". In *Studies in History and Philosophy of Science Part C: Studies in History and Philosophy of Biological and Biomedical Sciences* 44(2): pp. 225–236.

Schnell, U., F. Dijk, K. A. Sjollema, and B. N. Giepmans. 2012. "Immunolabeling artifacts and the need for live-cell imaging". *Nature Methods* 9: pp. 152–158.

Trogemann, G. (Ed.). 2010. *Code und Material—Exkursionen ins Undingliche*, first edition. Wien: Springer.

Xu, K., H. P. Babcock, and X. Zhuang. 2012. "Dual-objective STORM reveals three-dimensional filament organization in the actin cytoskeleton". *Nature Methods* 9.2: pp. 185–188.

7 Sonification and Audification as Means of Representing Data

Morten Søndergaard and Anette Vandsø

The research into scientific data representation is conventionally embedded in the techniques and technologies of visualization (cf. Tufte, 2001; Azzam & Evergreen, 2013, 2014). However, sound, and specifically non-speech audio, is used to represent data in many different fields including chaos theory, biomedicine, geoscience (such as seismology), physics (astrophysics and high-energy physics) neurology, genetics, and astrophysics seismology (cf. Hermann, Hunt & Neuhoff, 2011, 21; Supper 2012, 37–75). Well-known examples like ultrasound, interfaces for visual disabled people, seismic measurements, and the latest discovery of gravitational waves demonstrate that sound has the capacity of capturing the vicissitudes as well as impressive depth of detail in a rich variety of phenomena.

Lisa Gitelman, quoting Geoffrey Bowker, stresses that 'raw data is an oxymoron' because data is always already 'cooked' (Gitelman, 2013, 1). Her anthology investigates how data are 'variously "cooked" within the varied circumstances of their collection, storage, and transmission' (Gitelman, 2013, 3). Gitelman's point is that the chosen medium by which data is collected, stored, and transmitted is significant to how we can scientifically approach and analyze data as well as to the common, cultural ideas of what data is, or what "the world" is. Whereas the previous chapters have focused on visual representation, this chapter argues that auditory displays hold different possibilities and limitations than visual displays – it is a different way of "cooking", to use Gitelman's metaphor. In order to demonstrate the specifics of auditory data representation, this chapter presents key concepts such as "auditory display", "sonification", and "audification" through a number of specific examples. The chapter discusses how sound is constituent of our relation to data and ultimately to our ideas and understandings of the world, and it makes a distinction between the inherently energetic sonic representation of data and the more semiotic sign-based visual mode of representation.

Auditory Displays and Sonification

Even though the study of data representation tends to focus on visualization, the research into sonic representations has increased over the last 30 years, not least due to the establishment of the International Community for Auditory Displays (ICAD) in 1992. Since then, many substantial publications have scrutinized the field of non-speech audio data representation (i.e. Gregory Kramer's *Auditory Display: Sonification, Audification, and Auditory Interfaces*, 1994; editors Thomas Hermann, Andy Hunt, and John Neuhoff's *The Sonification Handbook* from 2011; and Alexandra Supper's PhD dissertation "Lobbying for the Ear" from 2012).

Auditory display is the general term that refers to the use of non-speech audio to display data. While the term "auditory display" encompasses a broad specter of hardware, software, and interactive processes, the specific process of turning data into sound is called *sonification* (Hermann, Hunt & Neuhoff, 2011, 1–2). Let us turn to an example from outer space: In August 2014, the European Space Agency (ESA) presented a spectacular new discovery made by its space orbiter Rosetta, which at that time was investigating the comet *67P/Churyumov-Gerasimenko* – a leftover from the birth of the solar system located near Jupiter's orbit. Rosetta's magnetometer, which is a highly sensitive instrument that measures the magnetic field in interplanetary space, had picked up an eerie "song" from the comet. ESA wrote about 'The Singing Comet' on its blog and posted a sound file that allowed the broader audience to hear the comet's song. The song was a sonification of data picked up by Rosetta.

According to the blog, this sonification was initially created as an operational tool for the scientific exploration of the comet (ESA, "Music of the Irregular Spheres" blogpost 19/12/14). The scientists hoped that certain patterns or structures that otherwise would not be detected could be revealed in the audible representation. The sonification also gained a lot of public attention – the "song" has been played over 5 million times. The sonification has proved to be a strong asset for ESA in communicating its current research project with the space orbiter Rosetta to a broader public.

Due to the text at ESA's website, the reader might think that the space orbiter has recorded sounds from the comet, but this is in fact not the case. In outer space, there are no sounds in the common understanding of the word, since sound needs a medium such as air to propagate. Other kinds of waves can, however, move in the comet's environment, for instance the magneto-acoustic waves picked up by the space orbiter's magnetometer. Accordingly, when the ESA writes about a "singing comet", it should be read as a metaphorical reference to the (non-audible) oscillations in the magnetic field around the comet.

The specific sounds we hear are instead created on a synthesizer by composer and sound engineer Manuel Senfft. The sonification is composed so that the frequency of the oscillations in the magnetic field determines the frequency of sound waves. Because the oscillations picked up by the magnetometer were only 40–50 millihertz, which is far below the human hearing range, they were increased by a factor of about 10.000 (ESA, "The Singing Comet" blogpost 11/11/14). The changes in the magnetic field are not a physical quantum changing over time but a more complex data set that can be described (ESA, "Music of the Irregular Spheres" blogpost 19/12/14). Senfft mapped these data parameters onto specific sonic parameters using advanced software so that changes in data result in changes in the sounds. For instance, we can hear small changes in pitch that directly reflect the changes in the frequencies of the oscillations. We can also hear changes in the stereo positioning and in the quality of the reverberation, which also reflect changes in the magnetic field.

The audible form of the sonifications determines what aspects of the data sets the scientists are able to hear. Senfft explains that he tried a few different ways of creating a sonification and rejected some of them because they were too 'exhausting to listen to' (ESA, "Behind the Scenes of 'The Singing Comet'" blogpost 19/12/14). But information regarding the mode of production also influences the general public's experience of sound. When the ESA writes that the sound comes from a comet, the referent seems to stick to the sound – we hear it as sound from a comet. Knowledge regarding the mode of production is thus relevant to and inseparable from the mode of reception.

The French literary theorist Gérard Genette calls the texts that surround artwork, such as the title of a book or the cover notes, 'paratexts' (1997, 1). Genette claims that the paratexts are not on the outside of the artwork, nor on the inside, because they condition our interpretative actions. The paratext is a threshold, he claims (1997, 1). Even though the singing comet is not an artwork in the narrow sense of the word, Genette's concept of paratexts is useful in order to understand how this sonification influences our understanding of outer space. The information at ESA's website functions as a paratext that conditions the interpretative actions we perform when listening to the sound. The text provides information about the nature of the sound, and after learning that the sound is from the comet, we cannot ignore this information.

It is not only the explanatory paratext at the website that gives the listener the impression that the sounds are coming from outer space; both the composition itself and its intertextual references to other sonic texts support this notion. This case of the singing comet reveals essential aspects concerning the use of sound as a means to represent data. It demonstrates how the act of representation is at the same time objective and in line with the positivistic scientific ideals, and artistic, and full of cultural, inter-textual references to musical styles and conventions, and it shows how sonifications often have several functions and purposes.

Senfft explains that he has deliberately attempted to create the illusion of an object moving in space. He has used different compositional strategies, such as stereo positioning, reverberation, and equalization, in order to create the mental image of an actual sounding object moving in space (ESA, "Behind the scenes of 'The Singing Comet'" blogpost 19/12/14, video 0:47). This gives the impression that the sounds are emitted from an object that moves and conveys the sense that the sonification is a signal. For instance, it sounds like the radio signals NASA's Cassini space orbiter picked up from Saturn (NASA, "Unlocking Saturn's Secrets" blogpost 25/07/05). The song of the comet also echoes something we know from science fiction features. In the blockbuster movie *Contact* from 1997, the main character Dr. Ellie Arroway (played by Jodie Foster) listens to signals from outer space, and some of these signals sound like the sound of the singing comet. As all other sonifications the singing comet draws, perhaps unintentionally, on intersubjective, cultural codes that are established through a large historical catalogue of sonic texts from various contexts, including sci-fi movies, documentary sound tracks, and electronic compositions. We simply cannot avoid activating our cultural encyclopedia in the act of listening (Eco, 1979).

When NASA made parts of its huge library of sounds from its space travels public, writer and blogger Dave Segal concluded with an ironic tone that "NASA Recordings Confirm That Deep Space Sounds Like '50s/'60s Avant-Garde Music" (Segal, blogpost 23/07/15). Segal mentions American composers Louis and Bebe Barron's soundtrack for the sci-fi thriller *Forbidden Planet* (1956) as an example of the aesthetics that seem to be repeated in the NASA sound files. It might, at first, seem unlikely that the sonic representation of data from space sounds like electronic music, but there is an explanation to it: Early electronic music of the 1950s and 1960s rose from experiments that included scientific technologies and methods – such as oscillators, EEGs (electrocardiographs), and seismographs – in various artistic set-ups often in collaborations between sound engineers, composers, and scientists. For instance, the American composer Alvin Lucier's *Music for Solo Performer* from 1965 is centered around the live sonification of the performer's brain waves. And in the German electronic composer Karlheinz Stockhausen's *Kontakte*, which he composed from 1958–1960, we are asked to listen to non-tonal, noisy and often unpredictable electronic signals *as* music. Electronic music was also used in sci-fi movies as a way of creating an atmosphere of outer space and of distant futures. In other words, the aesthetic characteristics of early electronic music that was used in *Forbidden Planets* and other sci-fi-movies, did not derive from the field of composed instrumental music understood as an isolated artistic practice but from a mixed practice where composers, sound engineers, and scientists together explored the possibilities of the audio technologies available in the mid-20th century.

All in all, the sounds of electronic signals in devices such as radios, Geiger counters, and oscillators are simultaneously "objective" sonifications of data and cultural signs enrolled in our common culture; they signify electronic music, and they are intimately connected to our shared imaginaries about the world, including outer space.

Not all sonifications sound as electronic music, and they do not necessarily require electronic instruments, though. On the contrary, conventional musical instruments or other real audio events are often used in order to convert data to sound. For instance, composer and data specialist Robert Alexander's sonification of solar winds made for NASA is a New Age–like, repetitive music made from non-electronic instruments including vocals, strings, and various percussion instruments.

Despite the very different musical expressions, both Senfft's and Alexander's sonifications are examples of so-called parameter mapping sonification where 'acoustic attributes of events are obtained by a "mapping" from data attribute values' (Hermann, Hunt & Neuhoff, 2011, 5; Grond & Berger 2011, 363–393). For instance, Alexander used a drumbeat to represent the rotation of the sun and a voice to represent the charge state of carbon atoms (*Phys news*, blog 26/02/10). Often a change in pitch is a parameter. In the case of the singing comet, the frequency of the oscillations directly determines the frequency, and thus the pitch, of the sounds. However, sometimes audio events such as footsteps can be used to represent data, or sometimes only one pitch is used, which is for instance the case of the ECG that measures the frequency of the heartbeat.

When we talk about sonification, the sounds we hear are a representation of data. Even though the sound is produced in an automated way, the sonification will vary from case to case, but once the composer or sound engineer has made a design for the sonification process, he or she will not have added further musical material. In this sense, the sounds are produced in an automated process. In many other scientific audio-visual representations, the sounds are not a reflection of data, and thus not a sonification, but a purely musical addition to the visuals.

Audification

In some sonifications, the data values directly define the audio signal. This subgroup is called *audifications*. This is the case with the electrocardiograms or seismographs that produce time-ordered sequential data streams (Hermann, Hunt & Neuhoff, 2011, 5). The following examples elaborate on the differences between audification and sonification.

A seismograph detects and records ground movement, including waves generated by earthquakes or volcanic eruptions. In early seismographs, the movements would be recorded graphically; later more complex seismographs translated movements into electrical signals that could be stored and also audified. In the 1960s, recordings were actually used for archiving seismological data, and it was common for the scientist to listen to the audified data. Since the frequency spectrum of Earth's movements is much lower than the human hearing range, which typically lies from 20 Hertz to 20 kiloHertz, the seismological waves have to be time-compressed, and often the dynamic range also needs to be reduced in order to be heard (Dombois, 2001, 1–2; Barass & Vickers, 2011, 147). Such audifications were used by scientists as a way to analyze data and by artists as an offset for creating music or sound art. For instance, Gordon Mumma's *Mographs* record series (1962) is based on specific seismic events (Kahn, 2013, 8).

Today both artists and scientists are still exploring this practice, often in interdisciplinary projects (cf. Dombois, 2001). The seismographic cases are interesting because they demonstrate that the sonified data is not necessarily a set of discrete, numerical data or a mathematical equation, and it is also not necessarily digital data. It might be a highly irregular waveform. We therefore use a broad definition of data such as the one found in *The Concise Oxford Dictionary of Mathematics* where "data" is defined as 'the observations gathered from an experiment, survey or observational study' (Clapham & Nicholson, 2009, "data"). Sometimes the data sets are first retrieved and stored in a non-audio medium and then sonified – as in the case with the singing comet – but the retrieved data can never be separated from its representation. And sometimes the primary data is sound. In the old electrical seismographs, the recording of an electrical signal onto magnetic tape that will store the data and allow us to listen to it using a tape recorder *is* the production of data.

Some of the sounds from NASA's library are audifications that are directly determined by data from outer space. This is the case with the radio emissions from Saturn mentioned earlier monitored by NASA's Cassini spacecraft initially detected in April 2003 (NASA, "Unlocking Saturn's Secrets" blogpost 25/07/05). The frequencies of the emissions have been lowered so 73 seconds correspond to 27 minutes, but otherwise the sound we hear is directly determined by the radio signals. Senfft and Alexander were working with a complex set of numerical data and assigning data values to sonic parameters. In comparison, the radio signals picked up by the Cassini spacecraft only had to be lowered in frequency, before being directly amplified and transmitted through a speaker.

Another example of an audification is media artist Anne Niemetz and scientist Andrew Pelling's art/science project *The Dark Side of the Cell* (2004) that allows the audience to listen to "singing cells" as it says on the website presentation of the project (Niemetz and Pelling, n.d.). They used an atomic force microscope (AFM) to register vibrations in cell membranes, mainly yeast cells. The cells were manipulated chemically in various ways, which created different vibrations in the cell walls. The frequencies of the vibrations were within the hearing range, but the amplitude was too small for the human ear to register them, so the electrical signals from the AFM were amplified in order to make them audible, but otherwise the sounds were directly created by the cell vibrations. Niemetz explains that:

> This process is not much different from reading data on an audio CD and amplifying the signals with a stereo system. The amplifying procedure intensifies the weak volume of the yeast cell sound but does not change the pitch or character of the sound itself.

(Niemetz 2004, 17)

In this case, the phenomenon represented is the movements of cell walls that are directly picked up by a very small needle and then transformed via a normal transducer into an electrical signal. This signal is amplified and turned into sound by a speaker. One type of energetic movements (oscillations in cell walls) is turned into another kind of movement (oscillations in air). Niemetz and Pelling have not had access to a set of numerically described or visual representation of data, which they have then sonified. In this sense, the process of audification is a way of representing a phenomenon directly – without data acting as an intermediator.

Sonic Representation

Just as visual representation creates an image of that which is being represented, the sonic data representation also involves a form of "imagery" insofar as the sounds create an idea or a mental image (Barass & Vickers, 2011, 154). However, using the auditory display holds different possibilities and limitations in comparison to the visual display.

Sound as a temporal phenomenon is useful for representing changes and repeating patterns in data. This is, for instance, one of the key points made by the researchers evaluating the sonifications of the solar winds (see Figure 5.2 in this volume for a visual representation of solar winds). Robert Alexander who does sonification of solar winds for NASA, states that with the music, they could hear patterns in the data that they would otherwise not have found and that the use of auditory analysis led to new insights 'regarding the source regions and ionic charge states of heavy ions in the solar wind, providing a very sensitive diagnostic of the electron temperature in the solar wind source region' (Alexander et al., 2014, 5260). Often the use of sonifications is done in combinations with visual surveying and quantitative analysis (Alexander et al., 2014, 5259).

In *The Sonification Handbook,* Neuhoff summarizes various experiments made with sonifications that use real audio events, for instance footsteps, instead of musical instruments. They conclude that not only can most listeners easily detect changes in audio events, as for instance the speed of paces, but they can also hear changes in the physical properties to the surface being walked on (Neuhoff, 2011, 78). They suggest that these cognitive mechanisms could be used for a simple sonification where the walking speed represents one variable in a multivariable data set, while the hardness of the surface represents another variable (Neuhoff, 2011, 79).

Professor of media and innovation Douglas Kahn (2013, 4) suggests the term "energetic" about art that is based on the transmission of energies (i.e. artworks that use electromagnetic waves, brainwaves, natural radio, etc.). When we look at the scientific and artistic uses of sonifications, they are often energetic in the sense that the sonifications represent energies, movements, waves, and not objects. They are transformations from one energetic form (electromagnetic waves, for instance) to another, namely sound – typically via electronic signals.

Kahn suggests that when thinking of energetic art, we need to replace our understanding of inscription, where semiotic signs are scripted onto a surface, with a focus on transmission:

> Inscriptive media precipitate phenomena onto surfaces (pages, scores, screens, memory devices, etc.) and are associated with recording and storage awaiting revivification, reproduction, repetition, and more storage. Transmissional media (in my usage) are inseparable from electricity and electromagnetism; they

differ from inscriptive media through basic physical states of energy (mechanics, electromagnetism) and are thus historically very recent when compared to the antiquity of inscriptive media.

(Kahn, 2013, 7)

Unlike musical compositions, sonifications are not notated in a score, and often they are listened to in real time – this is for instance the case with the Geiger counter where the focus is on transmission. When it comes to sonifications, the question of transmission versus inscription media is difficult, because sound in itself (waves in the air) cannot be stored. Consequently, the sounds of a sonification are not stored as sound but described by, say, grooves pressed in a vinyl record or zeroes and ones in a digital file. Still Kahn's point is valid and interesting because, first of all, the interpretative actions of the recipients are often not to decode signs but to hear changes in energies (frequency, for instance); and second, often the transmission of energies creates a particular relation between representation and the phenomenon. On a general level, we might say sonification functions as a *semiotic sign* with a form that refers back to its referent (Chapter 12 in this volume elaborates on sign-reference relations). With reference to Charles Sander Peirce, Chapter 2 in this volume accounted for the iconic and symbolic (diagrammatic) principles of representation, but a third principle offered by Peirce, indexicality, is of relevance in audification of waveforms. When seismic waves are represented by electromagnetic waves, a direct physical – that is, indexical – link exists between the sign (the audification) and the referent (the seismic waves). These kinds of sonifications allow us to experience the world, not in terms of separated objects – a comet, a cell, the sun, Jupiter, and so forth – but as movements or vibrational energies: the vibrations of cell membranes, the oscillations of a magnetic field, the outburst of solar wind, and the infrasonic, seismological movements of the Earth.

Even in the cases where the data is mapped onto arbitrary sounds, like when Alexander's sonification of solar winds by use of strings and vocals sounds like any other kind of New Age music, the sonification will often give its listener the impression of a world in movement. The aesthetic choice to use sound – which unfolds in time – and not a numerical or visual representation – allows the listener to experience the world as something that also unfolds in time, not as a stationary thing but something in constant flux. Compared to static images, sound has an inherent, unique ability to represent movement and flux and thus to represent these specific dimensions of the world.

Sonification as a Means to Communication, Investigation, and Aesthetic Experience

To what purpose do artists and researchers use auditory displays? As already mentioned, a sonic representation of data can be used in order to communicate the results and aims of current research projects to a broader public, or as part of an artistic practice. Auditory displays are also widely used in various gadgets and instruments, in particular in the field of medicine, where the ongoing retrieval and sonification of data is apt for surveillance of patient status. Finally, sonic representations are instruments for scientific exploration and examination (Azzam & Evergreen, 2013, 9). This is, for instance, the case with the sonifications of solar winds made for NASA. In this case, it is explicitly stated that 'the researchers' primary goal was to try to hear information that their eyes might have missed in solar wind speed and particle density data gathered by NASA's Advanced Composition Explorer satellite' (*Phys news* blog 26/02/10). But when NASA uploaded this material to platforms such as YouTube, Twitter, and Facebook, the sonification was also communicated in a way that appeals to different, non-scientific parties. Senfft's sonification was made in order for the ESA scientists to achieve a different approach to the retrieved data, but at the same time the composition has proved very useful in order to gain public attention.

Although many sonification practices are inherently interdisciplinary, in the sense that they involve close collaboration between scientists and composers, their specific presentation and circulation tends to be mainly in the fields of either science or art. The sonifications made for NASA and ESA are primarily produced and circulated within the social and professional field of science, while Mumma's *Mographs* records were produced and received by the art community. However, there are also examples of experiments that attempt to reach out to both areas. For instance, Florian Dombois's experiments with audification of seismological data is artistic research in the sense that the experiments are not directly used in a scientific context, but still its conclusions are specifically aimed for other seismologists (Dombois, 2001). Common to all the mentioned examples, whether they are art or science or in between, is that they allow us to experience a world we normally cannot sense, whether that world is outer space, inside our bodies, or the interior of Earth. Instead of distinguishing between art and science, it can be fruitful instead to distinguish between an aesthetic and a scientific mode of reception: In the scientific mode. we relate to the sounds we hear by analyzing it following the scientific methods; in the aesthetic mode the sound intrigues us in an undetermined manner.

References

Alexander, R., S. O'Modhrain, A. Roberts, J. A. Gilbert, and T. Zurbuchen. 2014. "The bird's ear view of space physics: Audification as a tool for the spectral analysis of time series data". *Journal of Geophysical Research* 119(7): pp. 5259–5271.

Azzam, T. and S. Evergreen (Eds.). 2013. *Data Visualization: Part I. New Directions for Evaluation*, p. 139. San Francisco: Wiley Press.

Azzam, T. and S. Evergreen (Eds.). 2014. *Data Visualization: Part II. New Directions for Evaluation*, p. 140. San Francisco: Wiley Press.

Barass, S. and P. Vickers. 2011. "Sonification Design and Aesthetics". In *The Sonification Handbook*, edited by T. Hermann, A. Hunt, and J. Neuhoff, pp. 145–172. Berlin: Logos Verlag.

Bennett, J. 2010. *Vibrant Matter. A Political Ecology of Things*. Durham: Duke University Press.

Clapham, C. and J. Nicholson. 2009. "Data". In *The Concise Oxford Dictionary of Mathematics*. New York: Oxford University Press. Cited 2016 April 8. www.oxfordreference.com/view/10.1093/acref/9780199235940.001.0001/acref-9780199235940. Accessed 15 Dec 2015.

Dombois, F. 2001. "Using Audification in Planetary Seimology". *Proceedings of the 2001 International Conference on Auditory Display*, pp. 227–230. Espoo, Finland: ICAD.

Eco, U. 1979. *The Role of the Reader: Explorations in the Semiotics of Texts*. Bloomington-London: Indiana University Press.

ESA. 2014. "The Singing Comet". *ESA—Rosetta Blog*, November 11. http://blogs.esa.int/rosetta/2014/11/11/the-singing-comet/. Accessed 15 Dec 2015.

ESA. 2014. "Behind the Scenes of 'The Singing Comet'". *ESA—Rosetta Blog*, December 19. http://blogs.esa.int/rosetta/2014/12/19/behind-the-scenes-of-the-singing-comet/. Accessed 15 Dec 2015.

ESA. 2014. "Music of the Irregular Spheres". *ESA—Rosetta Blog*, December 19. http://blogs. esa.int/rosetta/2014/12/19/music-of-the-irregular-spheres/. Accessed 15 Dec 2015.

Genette, G. 1997. *Paratext—Thresholds of Interpretation*. Cambridge: Cambridge University Press.

Gitelman, L. 2013. *"Raw Data" is an Oxymoron*. Cambridge: MIT Press.

Grond, F. and J. Berger. 2011. "Parameter Mapping Sonification". In *The Sonification Handbook,* edited by T. Hermann, A. Hunt, and J. Neuhoff, pp. 363–396. Berlin: Logos Verlag.

Hermann, T., A. Hunt, and J. Neuhoff. 2011. *The Sonification Handbook*. Berlin: Logos Verlag.

Kahn, D. 2013. *Earth Sound. Earth Signal. Energies and Earth Magnitudes in the Arts*. California: University of California Press.

Kittler, F. 1999. *Gramophone, Film Typewriter*. Stanford: Stanford University Press.

Kramer, G. 1994. *Auditory Display: Sonification, Audification, and Auditory Interfaces*. Reading: Addison-Wesley.

Levin, J. 2016. *Black Hole Blues and Other Songs from Outer Space*. London: Bodley Head.

Morton, T. 2013. *Hyperobjects. Philosophy and Ecology After the End of the World*. Minneapolis: University of Minnesota Press.

NASA/Robert Alexander. "Sonification of Solar Winds". YouTube Video. www.youtube.com/watch?v=kryCbfRJCyk. Accessed 15 Dec 2015.

NASA/Cassini. 2005. "Unlocking Saturn's Secrets". *NASA Blog*, July 25. www.nasa.gov/mission_pages/cassini/multimedia/pia07966.html. Accessed 15 Dec 2015.

Neuhoff, J. 2011. "Perception, Cognition and Action in Auditory Displays". In *The Sonification Handbook,* edited by T. Hermann, A. Hunt, and J. Neuhoff, pp. 64–85. Berlin: Logos Verlag.

Niemetz, A. 2004. "Singing Cells, Art, Science and the Noise in Between". Unpublished master thesis, University of California. Avaliable online: http://users.design.ucla.edu/~aniemetz/Niemetz_Thesis2004.pdf. Accessed 15 Dec 2015.

Niemetz, A and A. Pelling. n.d. "The Dark Side of the Cell". Cited 2014 December 20. www.darksideofcell.info/singingcell.html. Accessed 15 Dec 2015.

Phys news. 2010. "Scientists Listen to the Sun in New Sonification Project". *Phys.org*, February 26. http://phys.org/news/2010-02-scientists-sun-sonification.html. Accessed 15 Dec 2015.

Segal, D. 2015. "NASA Recordings Confirm That Deep Space Sounds Like '50s/'60s Avant-Garde Music". *Slog/The Stranger*, July 23. www.thestranger.com/blogs/slog/2015/07/23/22591471/nasa-recordings-confirm-that-deep-space-sounds-like-50s60s-avant-garde-music. Accessed 15 Dec 2015.

SOL. n.d. "Data of two Suncycles—1978–2000". Available online: www.sol-sol.de/htm/English_Frames/English_Frameset.htm. Accessed 15 Dec 2015.

Supper, A. 2012. "Lobbying for the Ear: The Public Fascination with and Academic Legitimacy of the Sonification of Scientific Data". PhD diss., Maastricht University.

Tufte, E. R. 2001. *The Visual Display of Quantitative Information*. Connecticut: Graphics Press.

8 Scientific Storytelling

Visualizing for Public Audiences

Djuke Veldhuis

In 1801 Humphry Davy demonstrated the latest advances in science at the Royal Institution, on Albemarle Street in London. Science was the hottest ticket in town. So much so that all those carriages clogging Albemarle Street created the necessity of introducing London's first one-way street (James, 2000, 6). Efforts to promote science and technology to the masses in the 19th century were not only successful but also highly profitable. The Great Exhibition made Prince Albert a profit of £186,000, an equivalent of tens of millions today (Williams, 2015). Six million people visited the Great Exhibition in London between May and October 1851, an equivalent third of Britain's population at the time and substantially more than the 3 million visitors the Science Museum in London welcomes each year today.

Davy's success as a public figure is only in part explained by his success as a chemist. More so than his discoveries, such as sodium and potassium, it is noteworthy that he engaged all the senses when communicating science to the broader public. Laughing gas would fill the Royal Institution, and explosions reverberated through its doors. Like children listening to an elder spinning fairy tales, so the audience was rapt with attention listening to stories of science. Similarly, the Great Exhibition was hands-on, from electric telegraphs and microscopes to air pumps and barometers (Forgan, 2000). For readers who feel that these examples are biased toward the "flash-bang-zoom" of science, here is an example from the field of statistics. Perhaps not a subject that typically conjures up excitement, nevertheless, it is from this field that the "Jedi-master of data visualization" and one of the world's 100 most influential people in 2012 hails (Provost, 2013). Instead of static histograms, Hans Rosling weaves statistics into a brightly colored tapestry of stories on health, wealth, and population using dynamic, animated data visualizations but also boxes, teacups, and toy bricks (TED, 2016). Although the value of the so-called "experience economy" has been questioned, evidence from psychology to pedagogy illustrates the value of engaging body and mind in an interactive manner and one that taps into the human predilection for narrative frameworks (Shams & Seitz, 2008; Dahlstrom, 2014; Boyd, 2001).

The central question explored in this chapter then, is: What strategies are used today by visualization practitioners to elicit the same wonder, engagement and curiosity in our 21st-century viewers as Davy and The Great Exhibition were able to? And how does engaging science communication simultaneously ensure accurate, accessible, and appropriate representation of data? This chapter investigates ways to develop and display storytelling visualizations of science aimed at broad (non-specialist) audiences in settings ranging from science festivals and museums to print and digital media.

Scientific Storytelling

The rate of data acquisition today is unparalleled. As demonstrated in the previous chapters, visualization – and sonification – is now actively required for exploration and analyses in addition to its role in presentation and communication. The Sanger Institute – one of the premier centers for genomic discoveries – for example, has 22 petabytes of DNA sequence data storage alone. One petabyte equals 1 million gigabytes – that is enough to store the DNA of the entire population of the United States and then clone them, twice (McKenna, 2013). But this growth in data does not necessarily mean a corresponding increase in useful information (Johnson et al., 2006). The challenge is to take wealth of information and turn it into a meaningful scientific story.

To complement this phenomenal rate in scientific data acquisition and indeed to ensure scientists can actually benefit from this data deluge rather than drown in it, hundreds of computer-based visualization tools have hit the market (O'Donoghue et al. 2010, 52). This has aided large-scale data analyses. However, identifying what tools are most appropriate for investigative research purposes versus those for communicating the results of research to a broader audience is by no means straightforward. In research and industry, many processes depend on people using and translating complicated representations of information (Dede et al., 1999, 282). Meanwhile, public audiences are asked to vote on science policy issues from carbon dioxide emission to genetically modified crops – areas governed by complex multivariate statistics, not easily visualized even for specialists. Sometimes of course, visualizations may serve both purposes simultaneously. Generally, the communication of results to a wider audience often comes as an afterthought to the research.

Effective visualizations are so powerful because they take advantage of our perceptual and cognitive systems, which are biologically predisposed to detect patterns and make spatial inferences and decisions. However, human information-processing capabilities, both visual and cognitive, are limited and systematically biased (Johnson et al., 2006).

If we want to generate the same excitement in our viewers as Davy was able to do, we do well to go back to the basics of telling a story. Why? Because we organize our experiences and memories in the form of stories (Bruner, 1991, 4). Evidence suggests that the narrative structure of a message not only affects how the message is processed and recalled but also enhances memory and comprehension itself (Lang et al., 1995, 108; Haven, 2007, 118–119; Hardcastle, 2003). The role of narrative may even serve a biological function: to simulate potential situations to aid decision making, the production and scrutiny of counterfactuals to test and refine one's behavior (Gottschall, 2013; Scalise Sugiyama, 2001; Boyd, 2001). For example, many cultures use fairy tales, (origin) myths, and (religious) stories to teach children about social norms or ethics and potential dangers, to promote cooperation, and to suppress selfishness. An alternative hypothesis might be that the ability to tell stories, to convey packages of information in an accessible and entertaining way, is a trait under sexual selection. Good writers, musicians, and artists certainly seem to receive a lot of attention in society.

Indeed, it is almost impossible for people to not weave random, un-patterned information into a story. This is beautifully illustrated by a simple experiment conducted in the mid-1940s by Heider and Simmel (Figure 8.1). They created a short animated film, consisting only of a big motionless square that contains a "flap" that opens up. There are also circles, triangles, and squares of different sizes. These move randomly around, into and out of the square through the flap. After participants watched the clip,

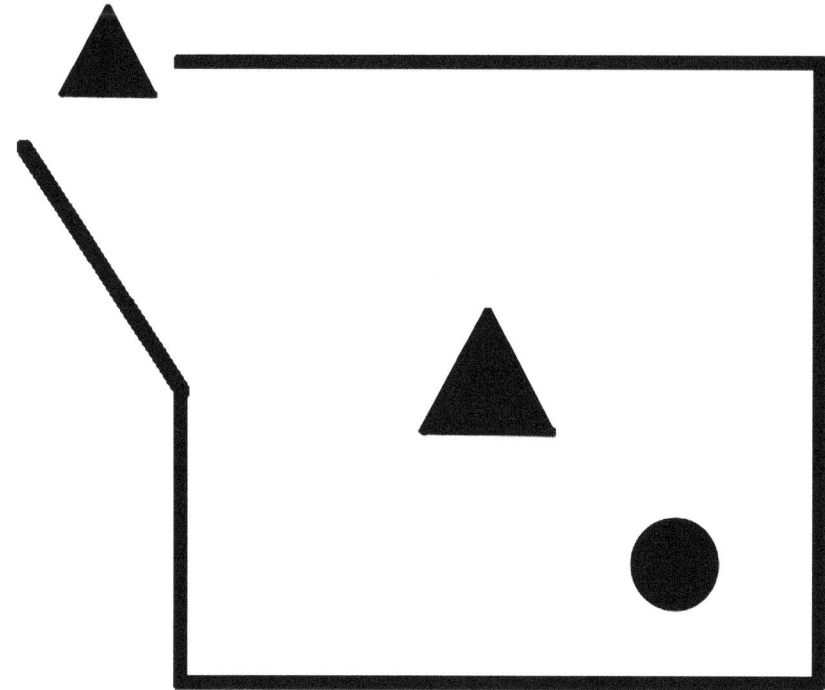

Figure 8.1 Redrawn from Heider and Simmel's 1944 experiment. This is readily available for viewing on YouTube: www.youtube.com/watch?v–VTNmLt7QX8E.

researchers asked them to describe what they saw. The result, replicated many times since then, was that only 3 out of 114 gave the reasonable answer: 'Geometric shapes moving around on a screen'. The rest described a rich soap opera of heroes and villains, doors opening and shutting, and good overcoming evil (Heider & Simmel, 1944).

Stories, more so than any list of facts or statistics, are a powerful means to transmit ideas and information in addition to supporting learning itself (Haven, 2007, 99; 101; Ragan, Mindt & Wittenberg-Lyles, 2005, 270). Setting, plot, and characters need to make an impression, pique the audience's curiosity, and leave them hungry for more. An engaging story takes a very predictable yet effective approach: It grabs our attention and introduces a crisis, which typically escalates, before a discovery catapults a protagonist into a cascade of change.

Pacing is particularly important when it comes to scientific storytelling. Any given pace may feel fast or slow depending on the audience's attention span and their familiarity with the subject matter and prior exposure to scientific data visualizations (e.g. graphs, line charts, maps; Ma et al., 2012, 12; Haven, 2007, 89–122). Familiarity with specific scientific imagery differs even within the sciences – as demonstrated by Chapter 6, this volume, in the account of physicists' and biologists' different pictorial cultures. Therefore, it is important that the pace of a good scientific story takes time to unfold and its pacing matches the audience's ability to follow. The forms these stories can take are infinitely flexible and broad in scope from "verbal designs" (public speaking) to "visual designs" (see Chapter 3, this volume).

Framing data as a narrative makes it more interesting and memorable (Ma et al., 2012, 13; Gusfield, 1976; Polletta et al., 2011, 122–123; Haven, 2007, 118–119) and allows researchers to guide the viewer. This is because we have two types of memory, episodic (used for remembering sequences of events) and semantic (used for remembering disconnected facts), and by presenting themselves as narratives, visualizations tap into episodic memory and establish themselves as cohesive entities. So how do researchers transform the rapidly growing number of scientific visualizations into meaningful experiences for the public? Before we can consider the use of narrative, it is pertinent to look briefly at the different types of data that researchers work with and how we may consciously, and indeed sometimes unconsciously, modify them.

Cooking Data: Raw, Smoked, or Baked?

The average reader, radio listener, or museum visitor will probably understand what a graph is meant to do – display information in (some sort of) a connected way – but will find it challenging to identify whether that graph is true to the original data. Typically, visitors are not given access to the raw data; nor should they if the objective is to let the "pictures speak" in lieu of data. Similarly, even if the viewer is presented with a visualization true to the original data, graphs, charts, and maps may all distort data on the basis of: angles (e.g. Mercator map); improper scaling (e.g. by exaggerating squares ratios); truncation (e.g. the y axis does not start at 0); or unfair statistical paradigms (e.g. using percentages where sample sizes are small). Distortion may also occur if loaded or leading words in legends – along with missing categories – prime the reader in one or another direction. Nevertheless, the scientist can make a conscious decision as to whether he or she represents the data raw, smoked, or baked.

Served "raw", an original data set is presented without modification, data point by data point. It is possible, for example, to search the complete human genome and compare it to sequences from other vertebrates ("Ensembl Project"). "Smoked" data has been modified and grouped into some sort of logical sequence. Continuing with genomic data as example, the U.S. National Library of Medicine provides digital tools that search for and identify specific portions of human genome in its repository of gene databases ("NCBI Human Genome Project"), which have been pre-categorized and ordered according to parts in the human genome. Users can request graphs showing relationships between different sections of the genome. "Baked" data has not only been modified and represented after data consolidation, but also explicit interpretations of that data have been made for the viewer. For example, the company *23andMe* will sequence your unique genome and, on the basis of genetic markers, extrapolate the specific health risks you may personally be more susceptible to based on population level genetic data (23andMe, 2016). In this example, only the "baked" data is intelligible to non-professionals.

Raw data in and of itself is of questionable value to laymen. Situations when scientists present data raw for public audiences (or professionals for that matter) are rare as that wealth of information necessarily creates a poverty of attention (Johnson et al., 2006). It is generally too hard to see overall patterns in a list of numbers or letters; however, in some cases, raw data use is suitable, for example when plotting patterns on geographic maps, floor plans, weather patterns, etc. (see Chapter 3, this volume). Generally, pre-visualization data modifications will involve a combination of baking and cooking. Discussion as to the amount of cooking or recipes to follow when it comes to visualization for public audiences is beyond the scope of this chapter. However, a useful starting point is Schneiderman's (1996, 2) *Visual Information*

Seeking Mantra: 'Overview first, zoom and filter, then details on demand'. Careful consideration of the type of data most suitable to get the message across in an exciting yet objective manner provides the bedrock of the story and will also influence the amount of control given to viewers to explore and interact with the data.

Turning Data into Visual Stories

A visualization becomes a visual story when it is linked by a set of story nodes, which represent important stages (key milestones or changes in scenery/evidence/tone etc.) of the visualization build-up process (Wohlfart & Hauser, 2007). Just like in a book, individual story nodes are connected through story transitions, which smooth over any changes in narrative or visuals that might confuse or distract a viewer. They essentially provide guidance as to changes in plot, setting, or characters.

Here, we discuss several practical examples using highly diverse formats but all constructed within the bounds of a (visual) scientific narrative. Our first stop is the hall of human origins at the Smithsonian Museum of Natural History. Here, an ingenious interactive display leads the visitor through vast amounts of evolutionary, climatic, and geological data. Next, we explore the rise of quick-fire, accessible science storytelling without the academic leash as we enter the world of (science) festivals, cafés, and (digital) edutainment.

Smithsonian Museum of Natural History: Human Evolution Exploration Booths

The David H. Koch Hall of Human Origins at the Smithsonian Museum of Natural History in Washington, D.C., takes a great effort in introducing a (scientific) plot, setting, and characters – in this case, the world in which our human ancestors evolved (Figure 8.2).

Figure 8.2 Meet your ancestors. 76 fossil skulls from 15 species of early human.
Image credit: NHB2010-01851 The David H. Koch Hall of Human Origins, National Museum of Natural History, Smithsonian Institution. Photo by Chip Clark.

Upon entering the hall, visitors travel back through time through a time tunnel depicting life and environmental change over the past 6 million years. They are then encouraged to enter a series of booths that envelop them and re-create actual archaeological field sites. In doing so, the Smithsonian attempts to create a memorable setting and to make the viewer an active part of the plot (to uncover the secrets behind human evolution). Talking walls and animation are overlain and integrated with three-dimensional replicas of fossils embedded in an African lakebed, which the visitors can touch, assessing the textures of different bones and soil substrates. Speakers filter in 360° audio of buzzing insects and wind sweeping across this African plain, which blocks out the general din and buzz of visitors in the main exhibition area. Altogether this provides an environment for young and old to explore this panorama, assess the evidence and complete "their" scientific task, for example, separating the human fossils from those of other mammals.

The opportunity to layer information can be achieved by using aforementioned story nodes; they enable "jumping off" points for interactive exploration (Ma et al., 2012; Brambilla et al., 2012, 76). For example, a basic short video is perfect for a visitor with limited time or superficial interest. For those wanting to delve deeper, the same display conveys much deeper information, for example, by providing additional narratives, such as an arrow tip embedded in the 3D lakebed display, which an observant visitor might have spotted and can choose to find out more about. Much like so-called "Easter eggs" hidden in computer games, exhibits can also bring in additional characters, clues, and story elements to engage highly driven explorers. Such "interactive visual storytelling" can be very powerful, and it is worth briefly considering, using the Smithsonian Hall of Origins as example, how to incorporate this mechanism.

The Smithsonian wants to be as clear as possible with the information presented. The challenge is that the variability of personal, social, educational, and cultural backgrounds of the audience are likely to be vast. Storytelling by nature is not completely interactive. A prescribed message in a video gives the museum full control as to content. However, the same video may lose one viewer due to an overload of information or bore another on the basis of too few details. One option is to give the audience full control of all the information at hand, but this creates the risk that the audience does not see or comprehend the message that the museum attempts to convey at all. This is what is known as the narrative paradox: The moment you make your viewers interactive participants, they move from spectators to "spect-actors" that can divert the course of the story you are trying to tell (Ma et al., 2012). To tackle this problem, Wohlfart and Hauser have created a taxonomy, which encourages scientists to think about how much guidance and control is given to the audience (Figure 8.3, Plate 15).

The Smithsonian's human evolution exhibit takes advantage of a blended interaction scheme. Its viewers are generally encouraged to be active participants rather than passive observers. There are more traditional, non-interactive, non-guided mini videos for passive viewing combined alongside materials allowing content interaction (e.g. ability to touch and feel stone tools and bones). Another benefit of using this multi-node story approach is that with the freedom to interact with multiple lines and levels of data from different angles (e.g. 3D scans of different fossils, which can be rotated, zoomed, and filtered on a computer animation, in addition to real fossil skeletons and skulls), observers can explore, assess, and (dis)approve of the saliency of the evidence in front of them. This is likely to lead to more trust in the presented visualization message.

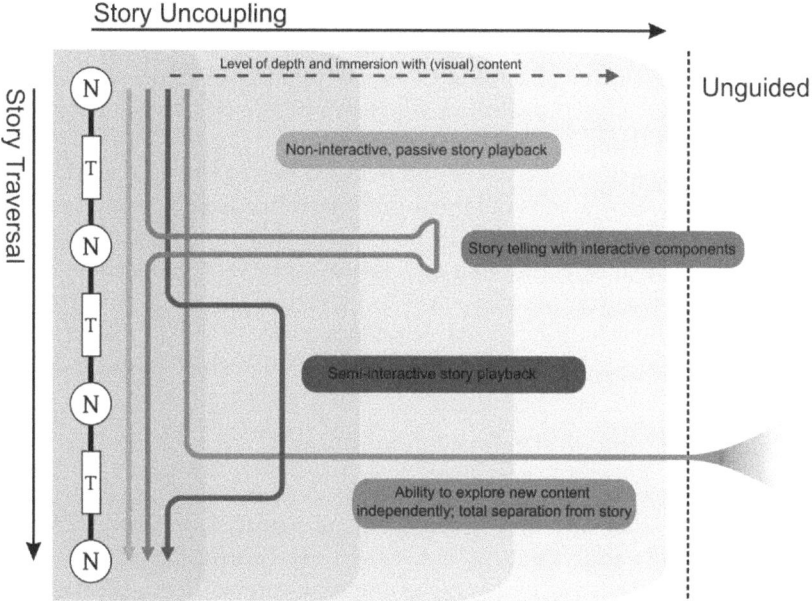

Figure 8.3 (See Color Plate 15) Possible storytelling and interaction schemes, modified from Wohlfart and Hauser (2007) by Joep Veldhuis. Non-interactive, directed, and passive story consumption: Here, the reader is fully guided by a story plot from start to finish e.g. in a video (**yellow**). Storytelling with interaction: At a story node, the audience has the ability to halt the story, temporarily take control, and independently explore e.g. rotate fossil bones 360° on a screen (**red**). Semi-interactive playback: Users can take control, not just for a brief excursion, but for an entire section, skipping certain story elements in the process e.g. rather than finding about early human ancestors, skip straight to the Neanderthals (**green**). Total uncoupling from the story where users are allowed to alter the scene and engage in total freedom (**blue**). *N*: story nodes; *T*: story transitions.

The Smithsonian's approach using these interactive booths not only conveys vast amounts of human evolutionary data, typically presented in academic literature as a dry series of comparative body and skull size measurements, but also allows it to tell the story of the fossil discoveries much more effectively. By placing visitors as protagonists in this story and giving them the responsibility as the paleoanthropologists responsible for this "discovery", the Smithsonian encourages visitors to engage in a proactive rather than passive manner. By taking on this proactive role, the audience is more likely to understand how researchers find, use, and interpret evidence to shape their understanding of human evolution. It turns on its head the dry and predictably faded panel displays typically located at the start of an exhibition hall, chronicling inventors, explorers, or scientists on monochrome photographs accompanied with equally stale bibliographic information. By the time the "real" characters are introduced in this Smithsonian exhibit, visitors are already primed to identify with their scientific prowess having spent time making the discoveries and untangling the evidence they, too, were confronted with.

Genetics, (paleo-)anthropology, paleontology, primatology, archaeology, zoology, and geology are just some of the subject areas one of these individual booths open up.

The combination of this interactive scientific narrative, alongside evidence in more traditional museum format, for example, the imposing floating fossil skull wall – serving simultaneously as an aesthetic installation for those not enamored by natural history – or life-size reconstructions of early humans by paleo-artists, all come together to create stories within stories, which are highly effective in hooking visitors into at least some of them.

Visitors can select, skip, and cross in and out of different story lines. Although some information may be lost, a self-contained mini-narrative is conveyed in each section. Unlike more traditional chronological linear exhibit, this provides the visitors with the freedom to choose their scientific journey. Taken together, the multisensory learning environment with visualizations steeped in a narrative provide a powerful means of conveying information using established techniques shown to assist learning and memory formation (Johnson et al., 2006, 7; Carroll, 1993, 304; Green & Bavelier, 2008).

From Festivals and Cafés to (Digital) Edutainment

Scientific storytelling has become a new tradition at museums, cafés, and festivals across the world. What arguably started out as exercises to improve researchers' communication skills and talent-scout for the next generation of Carl Sagans and David Attenboroughs has led to a wave of popular competitions, events, and online channels that encourage quick-fire, accessible science that is fun and educational. One of the oldest is FameLab, a global science communication competition sponsored by reputable scientific and cultural organizations (incl. NASA, British Council), which sees thousands of researchers compete each year. Using only their scientific storytelling wit, humor, and handmade props (PowerPoint is strictly forbidden), the competitors have just three minutes to present a scientific concept to a general public audience. Following in the footsteps of popular television talent shows, participants are under the scrutiny of judges and scored on "content, clarity, and charisma". FameLab takes place at public venues around the globe: in museums (where "live science" or "science open" entertainment nights are increasingly common), local theatres, pubs, cafés, or science festivals. Entry is free or a nominal fee as there is an onus on organizers to ensure research is accessible to anyone and everyone. Researchers receive communications training for free and exposure on a public, international stage, on television, radio, and online. Many similar endeavors exist both in more formal competitive settings (e.g. Three Minute Thesis or Alan Alda's Flame Challenge) and also informal events, which essentially encourage the public to come along to a local pub and ask researchers questions on their topic of choice over a drink (e.g. Pint of Science) and may include everything from science comedy to cabaret (e.g. Bright Club). These events epitomize the need for science stories since one cannot succeed in engaging broad audiences, of all backgrounds and ages, under time pressure, at public venues while speaking "academese" and rambling through a list of abstract facts full of technical jargon.

Both on stage and online, such educational entertainment or "edutainment" is ubiquitous. It has its opponents who feel that this apparent "sugar coating" of science to make it more palatable may undermine more serious communication efforts. However, given the millions watching YouTube science channels (e.g. Minute Physics, Brit Lab, Bill Nye the Science Guy, or ASAP Science), not to mention the rise of the "TED Talk generation", show that edutainment is here to stay. The stories they tell follow a standardized narrative format of "what does the data tell you about a daily problem or observation X?" And "how does it affect you?" Examples include: "Could we stop an asteroid?", "How much sleep do you need?", and "What if humans disappeared?" These stories of science are so successful because they either relate to the science of everyday life or pose questions of relevance and interest to people regardless of educational or cultural background. The content is accessible through bite-size information that is memorable and easily relayed to friends or family who will similarly be able to relate to the story even without an animated storyboard to guide them. These videos have become popular means of translating research into an accessible format as well as a means to attract public participation in scientific research.

More formal digital news portals are also moving in on an even more interactive version of such edutainment visualization stories. Like the "choose your own adventure" books, which encourage the reader to decide "what happens next", news portals sites are combining data, visualizations and interactive storytelling. For example, a recent news story based on data collected from refugees attempting to flee from Syria to Europe, gives readers the option to mimic that journey. Real-world data on the number of refugees, health risks, effects of gender and age as well as current political situations is presented before giving the viewer choices to survive "their" migration journey (BBC, "Syrian Journey: Choose Your Own Escape Route"). By placing the audience in full control at key points (story nodes), they decide how to act based on the data and information provided. This provides for a much more memorable experience than hearing of the constant refugee numbers in the news, which are difficult to contextualize in isolation.

The cases described here all demonstrate the effectiveness of (scientific) narrative as glue and vessel to effectively engage audiences and convey complex scientific data. With the narrative structure as a framework, the second half of this chapter considers the rise of citizen science: where visualization, research, outreach, and education meet.

A Few Pointers in the Use of Visualizations at (Science) Museums and Festivals

Because of the perceived self-evident nature of visual images, they are frequently repurposed, even copied and pasted, from scientific publications and handed to the media or museums for "illustrative" purposes. Little consideration is given to the viewer, who, without specialist training or fluency in the language of science, is likely to interpret the visualizations incorrectly. A good example of this is science center visitors' understanding of satellite images of sea surface temperature (from $0°C$ to $25°C$). Scientists typically report satellite data on a rainbow spectrum, regardless of the property being reported, and use purple to denote low values and red for high values. Aside from issues with such images being unintelligible for people with color-deficient vision, non-scientists cannot identify what is being presented. In this case, interpretation of sea surface temperature changes improved drastically when the color scheme moved from a gradated blue (at $0°C$) to bright red (at $25°C$; Phipps & Rowe, 2009, 313). This is of course the color scheme the public is exposed to on a daily basis in weather reports, whether in print, on television, or online.

Turney equates the job of translating science to "re-creating poetry" as it is translated from one language to another. Given the importance of cultural context, rhythm, and rhyme, this is more than just a direct translation of words alone. Culturally meaningful color and legend representations and measurement scales are key.

Beyond Representation: From Consumer to Producer, the Rise of Citizen Science

When volunteers, typically non-professional members of the public, help in the collections or in analyzing the data – whether in person, for example by recording weather patterns where they live, or online by examining images – we call it citizen science. Although volunteers, particularly amateur naturalists and astronomers, have been contributing to scientific research for centuries, the rise of the internet and applications to effectively crowd-source data has seen a proliferation of citizen science projects in the last 15 years. Several virtual citizen science projects – focused on images and visual data – effectively engage and educate public audiences by providing them with an active role as researchers who end up producing research results and output in their own right. In most stories and fairy tales, the basic narrative is one of good versus bad, a generic agonistic structure, which is systematically compelling and carries an adaptive function in evolutionary terms (Kjeldgaard-Christiansen, 2015). Citizen science harnesses these same instincts to motivate participants (Good & Su, 2011). It has led to a host of new discoveries via (virtual) visual data sets and adoption of gaming technologies. A host of new technologies, such as mobile applications (apps) and wireless sensor networks, show great promise in advancing citizen science (Newman et al., 2012)

An animated video, a mere one and a half minutes long, lays out a compelling story line of 'Sea hero quest: The first mobile game where *anyone* can help researchers fight dementia' (Spiers et al., 2016; BBC, "Mobile Game Sea Hero Quest 'Helps Dementia Research'"; italics by author):

> There was once a boy who lived by the sea. His father was a brave explorer. In faraway seas, they found incredible creatures. They captured them not with a hook but with a pen. They voyaged together until his father grew old, and then one by one, his memories were lost. Washed away by time and tide, the boy was now a man, and it was his turn to captain the ship and bring them back to his beloved Sea Hero!.
> (Spiers et al., 2016)

The viewer is challenged to take on the role of explorer and savior. The website asks the question as to whether the 3 billion hours spent gaming each week could be used for good and points out that playing *Sea Hero Quest* for just two minutes generates the same data that scientists would take five hours to collect in similar laboratory-based research. In other words, if 100,000 people play for just two minutes, they generate the equivalent of 50 years of lab research. A final stirring video, played with rousing music, shows the impact of dementia on people's lives and points out that there are no known cures or treatments and that scientists' lack of data is slowing down research. A powerful narrative encourages the player to sail a ship in search of sea creatures and test their navigational skills. The player's movements and choices are translated into visual snap shots – 2D heat maps – showing navigational decision making, which is anonymously relayed back to scientists for processing into binary data attributing scores to each development. In addition to generating visual data through gaming as in *Sea Hero Quest*, researchers can use citizen scientists to decode visual data.

"Classify a million images". That was the challenge facing astronomers in possession of the Sloan Digital Sky Survey containing 1 million galaxy images. Most galaxies come in two shapes: spiral or elliptical. Like our ability to recognize faces, distinguishing between these different galaxy shapes is easy for humans but impossible for a computer algorithm (Medeiros, 2014). The only way to do it is to look at each image one by one. Assuming a researcher was really efficient at classifying the images and spent only two minutes on each image and did that all day, every day, it would still take him or her nearly four years. Given the problem of observer error, this would only be the start, as you would need other researchers to replicate the findings of the original researcher. Faced with this conundrum, researchers launched *Galaxy Zoo* in 2007 (Lintott et al., 2011). This interactive web page displays images of galaxies that volunteers are then asked to classify as spiral or elliptical and to note any unknown objects or stars.

Within 24 hours of its launch, *Galaxy Zoo* was receiving 60,000 classifications per hour. After ten days, users from all over the world had submitted 8 million classifications. Just a month after its launch, a Dutch schoolteacher discovered a never-before-seen astronomical object that now carries his name (Medeiros, 2014). Using humans as pattern recognition machines, researchers could process data quickly and accurately but also enable and reward serendipitous discoveries.

Just like our ability to recognize the shapes of galaxies, the apparent intuitive nature by which the human eye and brain can digest visual data can also be harnessed for esoteric and abstract problems. Take, for example, protein folding. Each protein has only one possible structure, and finding it is tough given millions of possible three-dimensional shapes. The makers of *Foldit* noted that 'Biochemists often tell you that a protein *looks* right or wrong' (Medeiros, 2014; italics by author). They wondered whether they could create a game where players manipulate protein structures, by dragging amino acids, according to a prescribed series of (biochemical) rules. The more stable structure created, the higher the score. Like *Galaxy Zoo*, the site provides the users feedback on their progress, lots of "behind the scenes" science that people can delve into; there are forums to discuss problems as well as contests, shareable scripts, and "recipes" that users can use to design new algorithms and a leader board where you can see each other's names and scores. *Foldit* sent users the challenge to discover the structure of the Mason-Pfizer monkey virus, a protein that leads to AIDS in rhesus monkeys. Something that biochemists had spent more than a decade trying to solve, a small group of players uncovered in three weeks (Cooper et al., 2008). Aside from sheer fun and enjoyment in playing such games to solve some of the world's hardest scientific problems, another motivation is the simple human desire to figure out where you rank (in the social hierarchy) compared to others on the leader board. Upon discovering the Mason-Pfizer protein structure, players were asked whether they wanted their individual names on a scientific paper; they declined. Instead they wanted the name of their group to be included on the paper "Foldit contenders" (Medeiros, 2014).

These examples from astronomy and biochemistry to physics and neuroscience illustrate the benefits of using volunteers to analyze and generate (visual) data. They are no longer passive consumers of information fed top-down from the "ivory tower" but immersed and contributing to the story of science. The gargantuan visual data sets generated in fields like astronomy, biology, and climate science present potentially greater challenges to the scientific community, because extracting useful data from them cannot always be achieved using a computer program. Citizen science is a highly effective tool to analyze (visual) data sets, produce data, and promote collaboration and education between researchers and the public.

Crowd-sourcing has gained a strong following not only with the general public but also with the scientific community. Amassing a loyal following of volunteers and maintaining people's interests as crowd-sourcing projects grow is a challenge. Scientists compete for eyes to tackle their (visual) data. As a result, mass online citizen science hubs have arisen. The visually simple, intuitive, and consistent interfaces allow volunteers to easily jump into research projects from across the globe and find educational resources, literature, discussion forums, and contestant leader boards all in one place. For example, *Galaxy Zoo* is now part of *Zooniverse*: the world's largest and most popular platform for people-powered research with more than a million registered users, known as "Zooites" (Zooniverse, 2009; Hall, 2014). It provides scientific story lines for audiences of all ages and backgrounds. For instance, "Penguin Watch", which asks volunteers to count penguins from colonies across Antarctica ("Penguin Watch," 2016), is particularly popular in primary schools and arguably involves the world's youngest Antarctic researchers (Gill, 2016). By contrast, projects such as the "Higgs Hunter" (where volunteers assess images from the Large Hadron Collider at CERN for particle decay) or "Science Gossip" (where volunteers classify drawings and maps locked away on the pages of Victorian periodicals) have relatively complex themes and classification tasks, which are likely to appeal to adult audiences. Heavy social media engagement and the provision of free educational resources (e.g. the "ZooTeach" repository of lessons and resources for teachers) all encourage a continuing stream of players to feed the *Zooniverse* people-powered research machine.

Crowd-sourcing (visual) data may be considered the reinvention of the scientific method as far as its potential impacts are concerned. By framing problems and tasks in an accessible story line, with intuitive and interactive visuals that provide constant feedback on task success or gaming prowess, scientists can tap into the same psychology that drives more than 700 million people worldwide to play games online (Diele, 2013). As the examples above show, the future of citizen science is also inextricably linked to emerging (gaming) technology and software. Big data is already being amassed via countless health and fitness apps monitoring everything from global sleep patterns (e.g. Walch, Cochran & Forger, 2016) to perception and memory (e.g. Brown et al., 2014). The continuing rise of smartphone ownership provides opportunities for puzzles and gaming anywhere, anytime, taps into our predilection for play and competition, and serves as a potential mine of citizen science opportunities. Research has become a truly interactive and organic web. The public can provide researchers with data, help with the analysis of data, and, assuming the researchers are doing their job right, ultimately hear about the results of their work in return, in the form of a (scientific) story.

When All is Said and Done…

Scientific storytelling using visualization, whether through animation, popular magazines, games, or in popular settings such as festivals or cafés, reaps many benefits. Especially since universities and researchers are all engaged in "brand building". The rise of open source science and publishing, increased collaborations with business and industry, alongside dwindling funding resources necessitate that researchers make their research accessible, applicable, and engaging beyond the ivory tower. In addition to reaching broader audiences, it also allows scientists to gauge the public response to new developments. This can help inform future research. It is all very well to have the ability to develop new technologies, but if the public is uninformed and skeptical and afraid of its implementation, the political will and funding is unlikely to follow (see e.g. NanOpinion, 2015).

It is possible to tell fun and engaging stories of science. It is up to scientists, artists, and curators to have the gumption to delve into those stories.

References

23andMe. 2016. www.23andme.com. Accessed 25 Sept 2016.

BBC. 2015. "Syrian journey: Choose your own escape route". *BBC News*. www.bbc.com/news/world-middle-east-32057601. Accessed 15 April 2017.

BBC. 2016. "Mobile Game Sea Hero Quest 'Helps Dementia Research'". *BBC News*. www.bbc.com/news/technology-36203674. Accessed 24 March 2017.

Boyd, B. 2001. "The origin of stories: Horton hears a who". *Philosophy and Literature* 25(2): pp. 197–214.

Brambilla, A., R. Carnecky, R. Peikert, I. Viola, and H. Hauser. 2012. "Illustrative Flow Visualization: State of the Art, Trends and Challenges". *Eurographics*.

Brown, H. R., P. Zeidman, P. Smittenaar, R. A. Adams, F. McNab, R. B. Rutledge, and R. J. Dolan. 2014. "Crowdsourcing for Cognitive Science—the Utility of Smartphones". *PloS one* 9(7): p. e100662.

Bruner, J. 1991. "The narrative construction of reality". *Critical Inquiry* 18(1): pp. 1–21.

Carroll, J. B. 1993. *Human Cognitive Abilities*. Cambridge: Cambridge University Press.

Cooper, S., A. Treuille, J. Barbero, Z. Popović, D. Baker, and D. Salesin. 2008. "Foldit: Solve Puzzles for Science". https://fold.it. Accessed 12 Dec 2016.

Dahlstrom, M. F. 2014. "Using narratives and storytelling to communicate science with nonexpert audiences". *Proceedings of the National Academy of Sciences* 111(Supplement_4): pp. 13614–13620.

Dede, C., M. C. Salzman, R. B. Loftin, and D. Sprague. 1999. "Multisensory immersion as a modeling environment for learning complex scientific concepts". In *Modeling and Simulation in Science and Mathematics Education*, edited by W. Feurzeig and N. Roberts. New York: Springer New York.

Diele, O. 2013. *State of Online Gaming Report*.Spilgames. Web. pp. 1–17.

Forgan, S. 2000. "A compendium of victorian culture". *Nature* 403(6770): p. 596.

Gill, V. 2016. "Are these Britain's youngest antarctic researchers?" *BBC Science and Environment News*. www.bbc.com/news/science-environment-35964863. Accessed 12 Dec 2016.

Good, B. M. and A. I. Su. 2011. "Games with a scientific purpose". *Genome Biology* 12(12): p. 135.

Gottschall, J. 2013. *The Storytelling Animal*. New York: Mariner Books.

Green, C. S. and D. Bavelier. 2008. "Exercising your brain: A review of human brain plasticity and training-induced learning". *Psychology and Aging* 23(4): pp. 692–701.

Gusfield, J. 1976. "The literary rhetoric of science: Comedy and pathos in drinking driver research". *American Sociological Review* 1976: pp. 16–34.

Hall, S. 2014. "Zooniverse Reaches One Million Volunteers". *Universe Today*. www.universetoday.com/109413/zooniverse-reaches-one-million-volunteers/. Accessed 12 Feb 2017.

Hardcastle, V. G. 2003. "The Development of Self". In *Narrative and Consciousness: Literature, Psychology and the Brain*, edited by G. D. Fireman, T. E. McVay, and O. J. Flanagan, pp. 37–52. New York: Oxford University Press.

Haven, K. F. 2007. *Story Proof: The Science Behind the Startling Power of Story*. Santa Barbara: Greenwood Publishing Group.

Heider, F. and M. Simmel. 1944. "An experimental study of apparent behavior". *The American Journal of Psychology* 57(2): pp. 243–259.

James, F. A. J. L. 2000. *Guides to the Royal Institution of Great Britain 1: History*. London: Roayl Institution of Great Britain: pp. 1–22.

Johnson, C., R. Moorhead, T. Munzner, H. Pfister, P. Rheingans, and T. S. Yoo. 2006. *NIH-NSF Visualization Research Challenges Report*. California: IEEE Press.

Kjeldgaard-Christiansen, J. 2015. "Evil origins : A Darwinian genealogy of the popcultural villain". *Evolutionary Behavioral Sciences* 10(2): pp. 109–122.

Lang, A., P. M. Sias, P. Chantrill, and J. A. Burek. 1995. "Tell me a story: Narrative elaboration and memory for television". *Communication Reports* 8(2): pp. 102–110.

Lintott, C., K. Schawinski, S. Bamford, A. Slosar, K. Land, D. Thomas, E. Edmondson, K. Masters, R. C. Nichol, M. J. Raddick, A. Szalay, D. Andressecu, P. Murray, and J. Vandenberg. 2011. "Galaxy Zoo 1: Data release of morphological classifications for nearly 900.000 galaxies". *Monthly Notices of the Royal Astronomical Society* 410(1): pp. 166–178.

Ma, K., I. Liao, J. Frazier, H. Hauser, and H. Kostis. 2012. "Scientific storytelling using visualization". *Visualization Viewpoints* 2012: pp. 12–19.

McKenna, B. 2013. "What does a petabyte look like?" *Computer Weekly*. www.computer-weekly.com/feature/What-does-a-petabyte-look-like. Accessed 13 Dec 2016.

Medeiros, J. 2014. "How Turning Science into a Game Rouses More Public Interest". *Wired.co.uk*. www.wired.co.uk/article/science-as-a-game. Accessed 20 Jan 2017.

NanOpinion. 2015. "NanOpinion: Nanotechnologies where should they take us?" *European Union's Framework Programme for Research and Development (FP7)*. http://nanopinion.archiv.zsi.at/. Accessed 12 Dec 2016.

"NCBI Human Genome Project". 2016. www.ncbi.nlm.nih.gov/projects/genome/guide/human/. Accessed 12 Dec 2016.

Newman, G., A. Wiggins, A. Crall, E. Graham, S. Newman, and K. Crowston. 2012. "The future of citizen science: Emerging technologies and shifting paradigms". *Frontiers in Ecology and the Environment* 10(6): pp. 298–304.

O'Donoghue, S. I., A. C. Gavin, N. Gehlenborg, D. S. Goodsell, J. K. Hériché, C. B. Nielsen, C. North, A. J. Olson, J. B. Procter, D. W. Shattuck, T. Walter, and B. Wong. 2010 "Visualizing biological data—now and in the future". *Nature Publishing Group* 7(3s): pp. S2–S4.

"Penguin Watch". Zooniverse project. www.penguinwatch.org/. Accessed 11 Dec 2016.

Phipps, M. and S. Rowe. 2009. "Seeing satellite data". *Public Understanding of Science* 19(3): pp. 311–321.

Polletta, F., P. C. Chen, B. G. Gardner, and A. Motes. 2011. "The sociology of storytelling". *Annual Review of Sociology* 37(1): pp. 109–130.

Provost, C. 2013. "Hans Rosling: The man who's making statistics cool". *The Guardian,* 17 May 2013.

Raddick, M. J., G. Bracey, P. L. Gay, C. J. Lintott, C. Cardamone, P. Murray, K. Schawinski, A. S. Szalay, and J. Vandenberg. 2013. "Galaxy zoo: Motivations of citizen scientists".*arXiv preprint arXiv:1303.6886 (2013).*

Ragan, S. L., T. Mindt, and E. Wittenberg-Lyles. 2005. "Narrative medicine and education in palliative care". *Narratives, Health, and Healing: Communication Theory, Research, and Practice* 2005: pp. 259–275.

Scalise Sugiyama, M. 2001. "Food, foragers, and folklore: The role of narrative in human subsistence". *Evolution and Human Behavior* 22(4): pp. 221–240.

Shams, L. and A. R. Seitz. 2008. "Benefits of multisensory learning". *Trends in Cognitive Sciences* 12(11): pp. 411–417.

Shneiderman, B. 1996. "The eyes have it: A task by data type taxonomy for information visualizations". *Visual Languages, 1996. Proceedings 1996 IEEE Symposium on Visual Languages*. Boulder, CO: pp. 336–343.

Spiers, H., M. Hornberger, V. Bohbot, R. Dalton, C. Hölscher, E. Manley, S. Sami, R. Silva, and J. Weiner. 2016. "Sea Hero Quest". www.seaheroquest.com/en. Accessed 20 Nov 2016.

TED. 2016. "TED Speakers: Hans Rosling." www.ted.com/speakers/hans_rosling. Accessed 12 Sept 2016.

Walch, O. J., A. Cochran, and D. B. Forger. 2016. "A global quantification of 'normal' sleep schedules using smartphone data". *Science Advances* 2(5): e1501705–e1501705.

Williams, N. 2015. "How the great exhibition of 1851 still influences science today". *The Guardian,* 28 Aug 2015.

Wohlfart, M. and H. Hauser. 2007. "Story telling for presentation in volume visualization". *Proceedings 2007 Eurographics/IEEE-VGTC Symposium on Visualization (EuroVis 07)*, Eurographics Assoc: pp. 91–98.

Zooniverse. 2009. "Zooniverse." www.zoonivers.org/. Accessed 25 Oct 2016.

9 Communicating Science – Aesthetic Choices in Publishing

Kelly Krause

Why do we publish science? Chiefly, to communicate scientific discovery to the research community and also to ensure knowledge is archived for future generations. Search the mission statement of most scientific journals and you will find that sentiment at the core. Some multidisciplinary journals, like *Nature* and *Science*, primarily serve their main audience of scientists but also aim to communicate research more broadly, to science policymakers and the educated lay public. Specialist journals take the opposite approach, with intense focus on a single discipline.

Visual communication in a scientific publishing context is by necessity vast and varied – ranging from figures in research papers, to scientific illustrations in analysis and news pieces, to abstract representations of research on the cover. Each type of visual has its own set of best practices informed by purpose and intended audience. Research figures present data and other visual evidence from an experimental setup and are created for specialists in the field – for example, a picture of rows of gels are commonly used as evidence in cell biology papers. Visuals in news and analysis pieces are frequently created to explain concepts to non-specialists and usually take the form of annotated diagrams, illustrations and "info-graphics". Along this spectrum, covers are designed to engage the widest possible audience, from specialists to the lay public.

Journal covers play a unique role in visual communication, often taking center stage (as can be seen at press conferences, where enlarged poster versions of the cover are not uncommon). While covers are often used in this way at events, the main objective of any type of cover – be it a journal or a book – is to interest the reader in the contents within. As such, covers are designed to maximize audience engagement with the journal – to delight, inform, enlighten, and entertain. This can be done in several ways, using novelty (showing something for the first time) or surprise (showing something familiar in a different way), or with humor, or beauty. This chapter provides an insight into some of the considerations and choices that the visual design department at *Nature* makes in its daily work, using examples from our own cover designs.

The previous chapter discusses the merits of storytelling as a technique for visual communication, which is relevant here, in a discussion of journal covers (Chapter 8, this volume). Indeed, the best journal covers tell a story summarizing the research accurately, without misleading elements, into one visual that is easily comprehended. As an example: *Nature* published a paper on how anthropogenic electromagnetic noise disrupts the magnetic compass of some migratory birds. Initially, the authors suggested that we simply show a photograph of a relevant species of bird on the cover. However, we wanted to tell the story more completely, by creating a single picture that included several elements of the story: the bird, the disturbance, and the source of that disturbance. We worked with the authors to obtain a visual of a relevant electromagnetic source (a radio tower) and combined that with a visual representation of the offending waves as a subtle background. In the foreground, we placed a European robin that had just taken flight, giving the impression that it was battling the waves. In the end, the core idea of the paper was translated into a single, dynamic image (Figure 9.1, Plate 16).

Figure 9.1 (See Color Plate 16) Anthropogenic electromagnetic noise disrupts magnetic compass orientation in a migratory bird. Storytelling with covers: The subject of the research paper featured on the cover is how anthropogenic electromagnetic noise disrupts the magnetic compass of migratory birds, like the European robin (pictured). We combined images of a specific sort of radio tower that creates the noise, electromagnetic waves, and a relevant bird species in flight to give the viewer a good idea of the story at a glance. © Macmillan Publishers Limited.

Aesthetics as Applied to Journal Covers

The word "aesthetics" comes from the Greek, originally referring to the notion of "perception" rather than "beauty" or "taste" (the latter is a definition popularized in the 18th century). Thinking of aesthetics in this way is useful for visualizing science, as aesthetic choices (such as form or color) turn data (like sets of numbers) into graphics or images, enabling them to be perceivable and thus understood (Wilkinson, 1999). The most fundamental part of the creative process is to translate data into a form that can be easily perceived and therefore communicated. While a journal cover should feed the senses, the end goal is not simply to create something beautiful or tasteful. It must on some level communicate an idea and tell a story.

Because covers are more abstract than other types of visuals in a journal (such as graphs and diagrams), few rules or best practices apply uniformly to all covers. Those that do exist, such as good composition and clarity, can be broadly interpreted, depending on the image and the story it is trying to tell. In this way, it could be said that a greater variety of aesthetic choices can be made when creating covers than for other images in a journal. A cover visual can take any form, providing space for artistic vision and interpretation, in contrast to figures like charts and graphs, where form and function are largely predetermined. This makes the cover an interesting case study in the role of aesthetic choices in communicating science.

Case Studies

Laniakea

This first case study illustrates a step-by-step process of transforming scientific data into an engaging cover image. In 2014, *Nature* published a research paper that created the first ever map of our own supercluster of galaxies. The supercluster was named Laniakea by the authors, from the Hawaiian words "lani", meaning heaven, and "akea", meaning spacious or imme asurable. The paper described a new way to define where one supercluster ends and another begins and included sophisticated visualizations such as 3D models and maps in the research (Tully et al., 2014, 71–73). Based on the wealth of images, we decided to explore the possibility of featuring the paper on the cover.

The first step to creating a cover is for the designer (usually an art director or art editor) to gain a firm understanding of the scientific findings of the paper as well as the overall story that is to be told. At this stage, the designer consults with the editor of the paper and/or the authors to learn the main findings of the paper, how this affects the field, and possible future applications of the research. During this process, the designer will also become grounded in the visual language of the specific science involved, with an examination of the images used in the paper.

In this instance, we discussed the research figures and extended data with Leslie Sage, the editor of the paper, to gain an understanding of what the visuals were showing and how they fit into the larger story of mapping a supercluster. There were several images set in 3D space (with x, y, and z axes), but we were particularly interested in one of the 2D extended data figures, as it gave the most accessible view of the supercluster's location (Figure 9.2A, Plate 17). The editor explained the key elements of the figure: The rainbow color scale indicates density (with high-density regions in green and red and

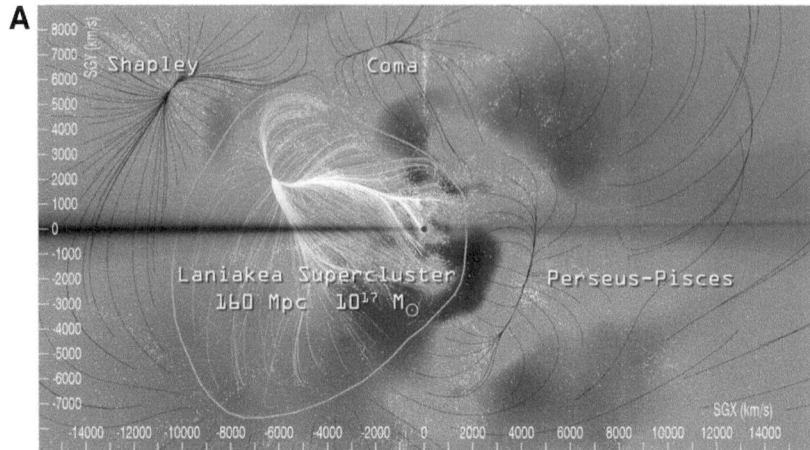

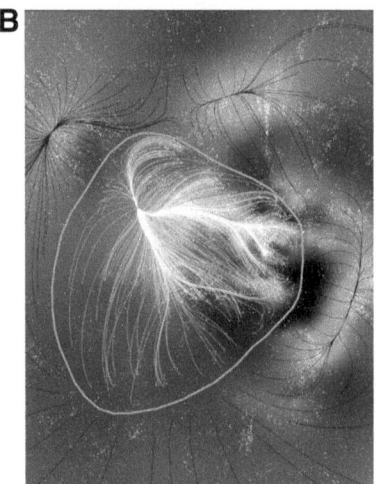

Figure 9.2 (See Color Plate 17) The Laniakea supercluster of galaxies. The making of a cover:
The author's original visualization of the supercluster (**A**); the author's second submission in response to our requests, to use as a reference for final artwork (**B**);
and the final art created by a specialist astronomical illustrator (**C**). © Macmillan
Publishers Limited.

low-density regions in blue); velocity flow streams are indicated by the blue and white
lines; and the orange band indicates the border of the Laniakea supercluster.

The next step is to develop an idea for a visual that translates the story into a
single cover image. This is an exploratory and highly collaborative process between
the designer and the editors and/or authors. In this case, we all agreed that the
2D map was the best representation of Laniakea for a cover, as the boundaries of
the supercluster could be clearly seen at a glance. But while the 2D image made a
very informative figure for a research paper, we felt that it needed modification in
order to appeal to a non-specialist audience. The visual techniques that represented
measurement, such as the rainbow-colored density scale, made the image useful as a
research figure but unrecognizable as map of outer space to anyone but specialists.
For a cover to communicate, it must be understood.

Working closely with authors Brent Tully and Daniel Pomarede, we requested a few modifications from them from which we would build a striking artist conception based on their data. We requested an amended version of the figure from Pomarede that shifted the rainbow density scale to a single dark gradient. We asked for this change for two reasons: chiefly, as mentioned, the bright colors made it difficult to imagine the supercluster as being set in outer space. Most people associate outer space with darkness, peppered by bits of light in the form of stars or planets reflecting their light. By following this convention – by changing the bright rainbow colors to a dark gradient – we were able to clearly set the scene in outer space. We also asked for the color change because the rainbow color scale is generally not the best technique to show quantitative data, as the shifts in color (say, from green to yellow) are not commensurate with change in value, creating misleading visuals (Gehlenborg & Wong 2012, 769). In short, the single dark gradient would still communicate the information while enabling a more recognizable outer-space setting for the supercluster (Figure 9.2B, Plate 17).

The final step is actually creating the visual. In many cases like this one, a specialist artist will be commissioned to create the final art. The designer will give the artist a brief that specifies what we want the image to show, how it should look, what to avoid, and any areas where the artist can be freer with his or her own interpretation.

We commissioned the Laniakea cover to Mark A. Garlick, an illustrator that specializes in astronomy. We briefed him on the paper's scientific findings and passed along the revised figure from the authors, plus our requests for modifications. Mark refined the density scale and merged it into a representative background that could be recognized as space and changed the Laniakea velocity flow streams (seen as purple lines on Figure 9.2B, Plate 17) to a warm glowing color that would be instantly recognized as light from the many galaxies in the cluster. We also removed the orange line that indicated the Laniakea border and replaced it with a more subtle approach, giving the supercluster a clear shape and with a visible border but in a layered, translucent style (Figure 9.2C, Plate 17).

Throughout the process of creating the final artwork, Mark worked with the authors to make sure his aesthetic choices were sound, such as the intensity of the glow from the velocity flow steams, and the subtle coloring of the density map. He was careful to ensure that inaccurate or misleading elements were not introduced into the visualization. This is an essential task when creating an artist conception based on scientific data.

The final cover resulted in an enlightening and engaging image – a visual that helps the viewer grasp not only the idea of a supercluster but also the viewer's own place in it via the orange 'You are here' text. All the aesthetic choices were calculated to enable perception, drawing on conventional visual elements – such as the glow of stars or the darkness of space – to help viewers comprehend something new.

In the above case study, the idea for visualizing research was fairly intuitive – using a spatial representation, or map, to help imagine our place in the universe. However, covers are not always this straightforward. There are times when the art is more abstract, employing metaphors to convey ideas, as can be seen in the next case study.

Positive ID

In 2015, *Nature* published a paper that provided the scientific community with framework for working with cultured cell lines. These lines, including many cancer cell lines, are essential tools not only in biological research but also in other areas of science. Unfortunately, it has been long known that many cell lines are contaminated, mislabeled, or incorrectly annotated. In the paper, the author Richard Neve and colleagues presented an analysis providing unambiguous authentication and annotation for more than 3,500 cell lines. Crucially, the paper also outlined simple measures to detect or avoid cross-contamination, presented a framework for cell line annotation, and provided a catalogue of synonymous cell lines. The paper is a key resource for the community, created with the aim of enabling researchers to eradicate misidentified lines and generate improved data (Yu et al. 2015, 307–311).

Due to the paper's large reach and importance to the community – and *Nature's* commitment to promoting scientific reproducibility, which the paper would positively impact – we decided to feature it on the cover. The first step, as always, is for the designer to gain an understanding of the paper and review the figures and other related imagery. In this instance, the figures were mostly graphs, tables, and charts.

The second step is to translate background knowledge gained in step one into an idea for the cover visual. Unlike the Laniakea supercluster paper discussed above, where a wealth of visuals in the paper supported the creation of an artist conception, this cover would have to take a different path: that of the visual metaphor.

The authors usefully supplied us with an idea – a cancer cell line that forms a question mark – to represent key ideas: the cancer cell lines being the subject of the research and the question mark representing the notion that many of the lines are misidentified and therefore unknown (Figure 9.3A). We liked how the visual metaphor told the story but thought that the question mark symbol as a representation of the paper was not quite right. After all, the paper is not merely outlining a known problem, but more important, it is presenting solutions. We decided to change the question mark to a tick mark (or check mark), a visual metaphor related to positive action.

With the concept decided, we went on to step three – creating the visual. We took the brief to specialist illustrators at AXS Biomedical Animation Studio, who carefully

Figure 9.3 (See Color Plate 18) A resource for cell line authentication, annotation, and quality control. Using visual metaphor: The author's submission (**A**); and the final artwork (**B**), a twist on the author's original question mark metaphor, rendered in 3D by a specialist medical illustrator. © Macmillan Publishers Limited.

constructed a lively scene of cancer cells in the form of a tick mark. The details are carefully rendered, blending accuracy with aesthetically pleasing composition and colors (Figure 9.3B, Plate 18).

The original submission employed a flat cellular style, as seen through a traditional microscope. We intentionally moved away from that toward a 3D object-based representation, as the original gave the impression of a static state. The illusion of depth in the 3D style allows the impression of movement, of cells caught in the act of gathering together, thus serving as a visual metaphor for the collective call to action set out in the paper.

The use of abstraction and metaphor can take many forms on a cover, such as illustration, discussed above, or photography, where things and places pictured often represent something broader. For example, a photo of a lone surviving frog tells the story of a disease that affects several populations of amphibians. Or a satellite image of a dying river represents a larger phenomenon.

Visual metaphors are clever and capture the imagination, but there are more direct ways of storytelling, such as the use of scientific imaging. At *Nature*, we frequently feature stunning images from the lab, such as microscopy or high-speed photography, as in the following case study.

A Work of Stiction

In 2016, *Nature* published a paper on stiction, or static friction, which is the force required to persuade an object to start sliding across a surface. It is technologically important in devices with moving parts but not well understood. The paper describes a model system (a switchable boron nitride surface) for the study of relationships between surface wetting, stiction, adhesion, and lubrication (Mertens et al. 2016, 676–679).

The authors sent in a cover submission (Figure 9.4A, Plate 19) illustrating how the shape of a liquid drop on a solid surface reveals details about the interaction between the

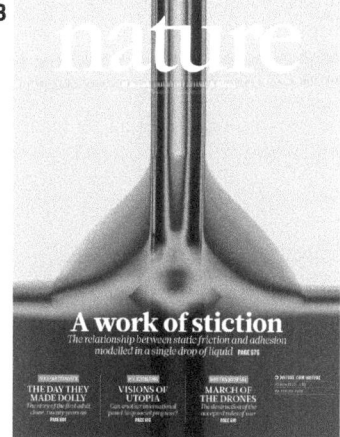

Figure 9.4 (See Color Plate 19) A liquid drop on a solid surface. Scientific imaging: The author's original submission (**A**); and the final cover (**B**), created by the author in response to our request for an image of a single drop. The final image uses 390 sequential 1 MB photographs, with bold use of color that is appropriate in the context of a cover.

two materials in contact. The image is a composite of a photograph and illustration, created in image-editing software, showing an illustration of a boron nitride surface superimposed on a photo of a lotus leaf, with a drop of liquid reflecting the surface pattern. While we sometimes employ this type of photo-illustration technique, we felt in this instance an image of a single drop of water on a plain surface (to demonstrate contact angles) might tell the story more directly – chiefly because the boron nitride surface on the leaf might possibly mislead readers when taken at a glance, as the paper itself did not actually involve plant surfaces.

We asked author Stijn Mertens if he had any visuals more focused on a single drop of water, and he came back with the excellent idea of using video material from the experiment to create a novel cover image (Figure 9.4B, Plate 19). To create the image, Mertens and team overlaid a number of recorded drop shapes – covering the spectrum from advancing (expanding) drops to receding (shrinking) drops – and rendered the resulting superimposed image in contrasting colors. The image uses 390 sequential 1 MB photographs of a drop in motion across the boron nitride surface. The result effectively illustrates the model in a single drop of liquid.

The most obvious aesthetic choice made by the authors was the strong use of color to depict the changes in contact angle as the drop expanded and shrank, helping illustrate motion in a static image. Color is a powerful tool and, when used wisely, can increase the impact of a visualization, as in this case. In the context of a cover, where abstraction is embraced, color can enhance a beautiful form or composition. In contrast, similar images in the paper were rendered in grayscale, as image integrity is the priority in that context. One should use caution when adding false color to grayscale images and avoid it entirely for research figures. As discussed in Chapter 3 of this volume, indiscriminate application of color can mislead, drawing the eye to artificially created areas of emphasis.

The above case studies are diverse in content and form, but all share the interplay between accuracy and aesthetic, detail and abstraction. Despite the variety of outcomes, the underlying creative process is more or less the same for all covers: Understand the content, research and brainstorm for the right idea, and execute the idea in visual form. During this process, various experts and stakeholders are consulted, options are weighed, and decisions are made. Journal covers are commercial products. The designer must produce something compelling on deadline every week. This forces questions of truth and beauty in a regular, concrete way. The cover of a scientific journal is, on a practical level, where theoretical constructs of aesthetics and real-world best practices of visual representation collide.

References

Gehlenborg, N. and B. Wong. 2012. "Points of view: Mapping quantitative data to color". *Nature Methods* 9: p. 769. doi:10.1038/nmeth.2134.

Mertens, S., A. Hemmi, S. Muff, O. Gröning, S. De Feyter, J. Osterwalder, and T. Greber. 2016. "Switching stiction and adhesion of a liquid on a solid". *Nature* 534: pp. 676–679. doi:10.1038/nature18275.

Tully, R. B., H. Courtois, Y. Hoffman, and D. Pomarède. 2014. "The Laniakea supercluster of galaxies". *Nature* 513: pp. 71–73. doi:10.1038/nature13674.

Wilkinson, L. 1999. *The Grammar of Graphics*. New York: Springer Publishing.

Yu, M., S. K. Selvaraj, M. Liang-Chu, S. Aghajani, M. Busse, J. Yuan, G. Lee, F. Peale, C. Klijn, R. Bourgon, J. S. Kaminker, and R. M. Neve. 2015. "A resource for cell line authentication, annotation and quality control". *Nature* 520: pp. 307–311. doi:10.1038/nature14397.

10 Ideas in Action

Using Animation to Cut through Complexity

Janet Rafner and Rikke Schmidt Kjærgaard

Graphic artists and information designers often represent data through animations, defined as the technique of presenting successive drawings, photographs, or models to create an illusion of movement when showing a film as a sequence. This is often done by applying a specific software code or algorithmic behavior to map the data onto a visual space or by using cross-platform 3D applications with pre-programmed features. Either way, the purpose is to create and explore idea-generating visual data representations. While the tools are widely used in the arts (e.g. in the creation of movies and computer games), they are progressing in the sciences as well. Some animations built from code have an element of interactivity, a very popular feature within learning and exhibition environments (e.g. Wilson et al., 2015).

Scientists use a wide range of traditional tools to visualize scientific data, and most of these are highly technical, computation-based, and scientifically advanced. They are designed to handle very complex data, and lots of it. Often, the software is associated with lab equipment and not intended to provide an aesthetic or artistic interpretation of the data but rather to concisely display the data so that the scientist is able to adjust and analyze it in terms of scientific value. Some tool designers have a mild interest in offering features to enhance the aesthetics or craft the subtleties of visualizations, but generally it is considered to be of lesser importance and value. However, with the increasing number of data designers and data artists, the possibilities and value of more advanced and aesthetically refined models for representing data are becoming more common in the sciences.

This chapter discusses the representational and aesthetic possibilities, advantages, and constraints that animation as a tool provides for data representation. It builds on previous chapters in this volume by extending ideas and concepts from general data representation to scientific animation and then to moving images, influenced by tools provided by the entertainment industry. Moving beyond current, or at least typical applications, this chapter also discusses what is and will be possible in scientific data and conceptual representations when using animation software. It looks at the possibility of extending software tools from the entertainment industry into active research tools for science and questions and considers the aesthetic implications thereof.

The Animation Process

To design and plan an animation, a clear aim or purpose needs to be defined. Typically, the purpose of the scientific animation is to convey and inform about the current understanding of the scientific topic. Animation techniques range from cartoons and

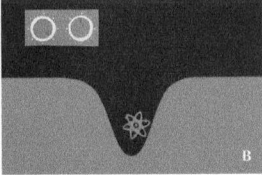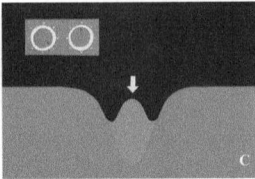

Figure 10.1 (See Color Plate 20) Still images from the stop-motion animation *How your gameplay is used to help build a quantum computer* directed by Janet Rafner and Pinja Haikka. **A.** How quantum computers execute multiple calculations simultaneously. **B.** An atom in a movable energy well. **C.** The same atom as a wave function showing its most likely position. The wave function is a more scientifically accurate description of an atom.

time-lapse to stop-motion and 3D animation. Typically, the first thing to do when starting an animation is to gather as much information as possible about the scientific subject, creating a variety of useful insights. The gathering of relevant knowledge is fundamental in creating a script and storyboard. The script defines the characters, the environment, and the all-important timing of the animation.

Cognitive studies have tried to establish what ingredients make for an efficient animation (Wouters, 2008). Animations may seek, for example, to illustrate a procedure, a natural phenomenon, or changes in data and its implications over time. And in each context, there will also be different priorities or audience characteristics. Different animation techniques and intended audiences allow for a variety of artistic license. Stop-motion animation, for example, can be used to make complex phenomena in science less intimidating. The medium is "constructed", and the viewer knows intuitively that the author has exercised control in crafting each frame, in effect becoming a storyteller (on the importance of storytelling see Chapter 8, this volume). The types of materials often used in stop-motion, such as paper cutouts, marker board drawings, and clay objects or figures, connote to the viewer, "Don't worry – we are going to make this as simple as possible and we want you to have fun, too". The viewer relaxes, understanding that his or her role is to be an audience – to watch the story. The genera of stop-motion gives the animator latitude, both technically and culturally, to readily shift back and forth between media types, to inject humor and metaphor, and to focus on the big-pictures concepts rather than precise representations of specific data.

The stop-motion animation in Figure 10.1 (Plate 20) was created by the ScienceAtHome research group, a large-scale crowd-sourcing project based at Aarhus University in Denmark. The research group developed a game called *Quantum Moves* that translates gameplay data into data that can be optimized and used in the experimental laboratory. The animation is used to answer the question, "What are quantum computers, and how does playing games help physicists in cutting edge research?" The video (4:24 minutes) provides a simple, concise, and fun presentation of an extremely complex topic (Link to video: www.youtube.com/watch?v=X24ombN09_k).

Capturing the Essence of the Intangible

Animation allows the artist to employ, but then go beyond, what can be photographed – to zoom to scales past the optical, illustrate a molecular structure in form and action, and then "pull back" to a simplified figurative illustration to show the broader interactions. This approach allows us to see, in one "take", what is otherwise impossible to see. For example, to start at the inter-cellular level of collagen fibroblasts, then zoom to a molecular level rendering of intracellular cytoskeletal structures, and then step back to a figurative animation of tensions between such structures, as contraction is induced by a protein. In other words, animation allows us to transcend scale, control time, escape the limitations of instrumentation, and illustrate interactions that are invisible yet fundamental to how natural processes work – and to craft it in a way that is both comprehensible and visually engaging. The temporal dimension allows for simultaneous audio or text narration, along with features such as moving pointers and time and scale notations that not only provide clarity to the content but also improve retention.

Visual representations of the unseen have led to a number of concerns about methodology, representational truth, and imaginative range and constraints. In the 1960s, molecular scientists used crude models built from wires and rods, and sophisticated illustrations were a rarity. The watercolors by scientific artist Irving Geis during these years, done from density maps and in close collaboration with scientists, constitutes the very beginning of volumetric 3D iterations of molecules and cell environments (Schmidt Kjærgaard, 2011). Moving from X-ray diffraction to computer-generated 3D models, molecular science has evolved significantly through the means of representational technology and made us rely on the machine to 'promise a non-human precision' (Kemp, 2006, 321). The increasing level of data complexity introduces new choices for science and data representation highly dependent of machines and of our conceptions. In molecular biology as well as modern physics, scientists constantly confront data that can only be "seen" though high-end technology, imagination, and visual representations. The graphic artist's and information designer's skills thus play a crucial role in shaping our understanding and beliefs.

Representational Features

Objectivity

The objectivity we demand in scientific "still images" is also demanded of moving pictures. Science illustrators and animators cope with the struggle between objectivity and accuracy to 'ensure the integrity of the object and subject' (Daston & Galison, 1992, 90). By illustrating scientific phenomena, they can publically promote research, as well as distribute information to the many versus having the knowledge remain limited to the few who actually conduct experimentation and research.

Ideally the goal of an objective portrayal eliminates the necessity of an illustrator to interpret, merely extrapolating information, while the desire for objectivity promotes 'painstaking care and exactitude, infinite patience, unflagging perseverance, preternatural sensory acuity' (Daston & Galison, 1992, 75). However, it is impossible to detach the subjective human eye from scientific discovery. The ability for humans to make scientific discoveries is inseparable from humans' ability to draw connections, make interpretations, consider problems from different perspectives, and harness the power of subjectivity. The progression of visualizations parallels technological advances, while the emphasis lies not with producing more results but better results with a focus on taming subjectivity. The skilled artist knows what to leave out, how to make subtle yet important variations, which perspective to choose, when to be naturalistic or precise – and when to be figurative or indistinct.

Selectivity

During a process of creating an animation for presenting scientific data or phenomena, many different facets will typically be simultaneously at play. The animator should pick a focus to avoid confusing the viewer. Consider the example of an architect's rendering of a proposed new campus building: The structure's form and position are precise, but the landscaping – the trees and shrubs – might appear as little more than fuzzy green cloud shapes. Of course, today, our near limitless computational power, fractal equations, and rich object libraries make it trivial to paste in completely photo-realistic plants, cars, street signs, and age and ethical diverse pedestrians – right down to their specific brands of sportswear. But the campus building itself would visually drown if the image was flooded with such details. Many graphic artists take great pride in creating images so flawless that you question whether it is not an actual photograph. We have come to value and be in awe of such is-it-real-or-is-it-computer-generated perfection. But in most cases, the far simpler rendering better achieves the goal of getting the viewer to focus on and contemplate the distinctive attributes of the new building.

The example described in Chapter 1 in this volume illustrates the necessity of selecting a single truth for a particular image. In this example, as well as in most molecular illustrations and animations based on crystallography and protein structure analysis, the orientation of each molecule is not realistic but shown by easily recognizable structures. A general and arbitrary orientation where parts of a molecule would "stick out of the plane" would not be recognizable and thus not fulfill the purpose of the scientific illustration.

Viewers who are unfamiliar with complex content will naturally have difficulty knowing which features to focus on. Thus, it is preferable that each step or phenomena should first be portrayed independently before being combined, with the relevance of each made clear. When multiple facets must be introduced and portrayed together, they should be directly related or supportive of a main concept. To a certain extent, this is just implementing a selective, step-by-step model, but it requires exercising discipline and not getting carried away by employing animation features that distract more than they inform. The potential for distracting artifacts can also be introduced by the very tools that help provide greater aesthetic control. For example, multiple (virtual) sources of light together with shadow and subtle color shifts are often crucial in enhancing realistic features. However, the premise of "realistic" is somewhat malleable when animating a scientific phenomenon that cannot be seen by the naked eye but rather only by use of technically advanced laboratory equipment. The implications of choices in color, size, motion, time frame, and light become much more delicate because these features implicitly become a concrete image that the viewer retains, not merely the abstraction they are trying to represent.

Quantum physics are an excellent example of scientific phenomena that cannot be seen by the naked eye – and has no well-accepted mode of representation. In fact, the question of representation itself often becomes the central topic of animations about quantum phenomena, such as the illustration of the dual wave-particle nature (Figure 10.2, Plate 21).

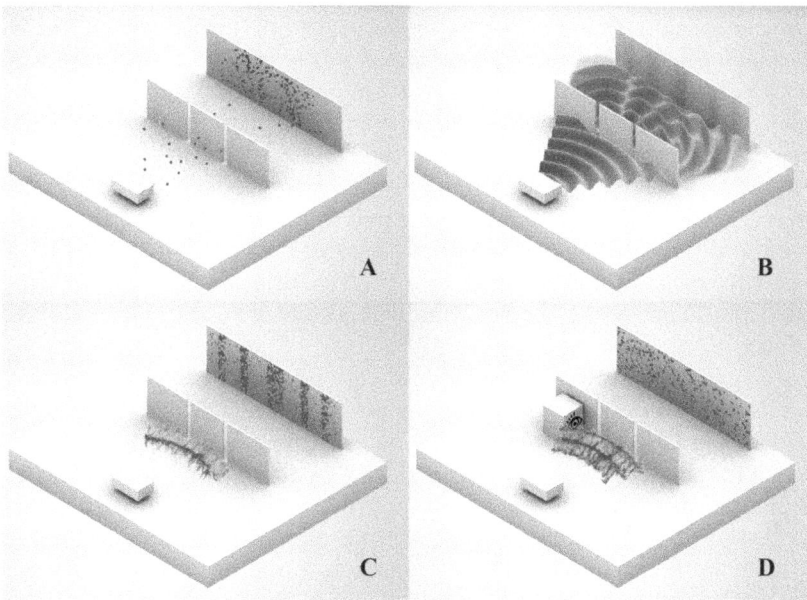

Figure 10.2 (See Color Plate 21) *Quantum Made Simple*, traditional animation. Screenshots of the double slit experiment animation, displaying **A.** particles, **B.** waves, **C.** quantum wave functions, **D.** quantum wave functions with an observer at one of the two slits (resulting in the destructive interference pattern). Full animation available at www.QuantumMadeSimple.com.

The collection of *Quantum Made Simple* animations, published by Professor Julien Bobroff, physics professor at the University of Paris Sud, seek to advance public and academic understanding of the forces at play in the nanoworld – forces of electromagnetic or quantum nature that bond together or manipulate atoms or make it possible to visualize them with atomic microscopes (Bobroff, Bouquet & Jutant; Figure 10.2, Plate 21). The animations last from one to two minutes and are accompanied by a brief subtitled legend. These processes captured in the *Quantum Made Simple* animations are immensely complex, and thus a lot of information needed to be removed in order to maintain the level comprehensibility for the intended audience – high-school students and teachers.

One way to control the flow of information and guide the eye of the viewer is through a chosen perspective. A combination of overview and zoomed-in perspective running simultaneously can be advantageous when focusing in on an element of a complex process. The *Quantum Made Simple* animations seen above are 'displayed in an isometric framework. The horizontal plane is the basis for the quantum phenomena while vertical planes display physical measurements or mathematical spaces (energy levels, periodic table, STM imaging, reciprocal space, etc.)' (Bobroff, Bouquet & Jutant). This helps to maintain scientific credibility even though 'the animations were not computed from exact simulations and do not represent a rigorous treatment of quantum physics. They were developed as an outreach easy-to-understand approximation of quantum phenomena' (Bobroff, Bouquet & Jutant). There remains a delicate give and take between the constraints of simulations and leaving freedom for innovative graphic design.

Complexity

The selective process that the animator needs to go through when planning the animation storyboard is very much connected to the level of complexity of the data. For example, to get a particular point across in a molecular animation, an illustrator might choose to omit the presence of water or other molecules and show only a subset of the molecular environment to highlight the active site. Using computer graphics to represent the molecular world demands cutting corners. In the animation by animator Drew Berry, shown in Figure 10.3A, we see only the part that is related to our specific narrative about DNA and how it is replicated inside the cell. A contrasting example is the much more realistic simulation by Sean R. McGuffee and Adrian H. Elcock from 2010, showing more than a thousand individual molecules diffusing, colliding, and interacting with each other (Figure 10.3B).

However, problems emerge in this representation as well: Each of these solid figures is a rigid body, an idealization without any deformation (as is also seen in the watercolor painting in Chapter 1, this volume), which is not how things work. It is often necessary and acceptable to reduce complexity by leaving out 'rapid movements in shots depicting entire proteins but not in shots focused on active site chemistry where thermal motion, bond stretching, and bending remain important factors' (McGill, 2008, 1129). An increasingly common approach to the challenge of depicting complexity while being selective is to start an animation in a broad context and low granularity and then zoom into the specific feature under study. However, this approach likely requires creating and sequencing together what amounts to multiple animations, each perhaps built from different development tools.

Color

While color is an artistic liberty of the animator, it has a deeper importance rooted in human cognition. Through the case of HIV, scientist and molecular animator Tim Seiber shows how aesthetics and choices of color and appearance play an essential part in debates concerning the human body and its microscopic parts. Culturally a magnified, graphically isolated, false-colored chemical abstraction, such as a virus molecule, separates the molecule from the human body (Seiber, 2014, 267). By adding color and creating a vulgarized environment, the animation produces an aesthetic

Figure 10.3 Molecular animations. **A.** Screenshot from an animation by animator Drew Berry from 2011. **B.** Screenshot from a molecular simulation by Sean R. McGuffee and Adrian H. Elcock from 2010.

doctrine much different from unprocessed data, usually presented as collection of numbers or micrographs.

Scientists look for patterns, and data designers and data artists are experienced at enhancing and putting these patterns forward in a wide range of different forms. The human mind responds positively to recognizing patterns, and thus, by using similar colors to illustrate similar concepts, the repetition will reinforce the understanding of the idea. For example, in *How your game play is used to help build a quantum computer* (Figure 10.1, Plate 20), all quantum mechanical elements (qbit, atom seen as an object, atom seen as a wave) are "color coded" as purple. This color consistency is especially beneficial when the pictures are moving very quickly or, as is in the case of the stop-motion animation, cutting between different representation models.

In the *Quantum Made Simple* animations (Figure 10.2, Plate 21), choosing red to represent the quantum matter helped 'to convey the difference between quantum objects, and standard waves or particles. In order to ensure consistency among animations, the same red matter was used for wave functions whatever the context' (Bobroff, Bouquet & Jutant). Without a rational comparison, reference point, or fundamental basis, concepts lack meaning. The human mind uses categories to organize the world and establish reference points for new information. Categories are determined based on salient relationships, however, they are not unitary or even permanent (McVittie, 2009).

Sound

In the pursuit of evoking a certain emotion in the viewer, sound is often an essential part of storytelling. Very basic sound effects can also directly support specific actions or visual effects in the animation – similar to the subtle "shutter" sound a smartphone might make when taking a picture, when two objects meet and connect in a "clink", or if an object accelerates and the pitch of a sound shifts higher. Background music or a sound/narrative track may be used in order to engage and captivate the audience and to set the tone of the animation (on sound as a representational feature in Chapter 7, this volume).

Picking or composing soundtracks is of course a specialty in its own right, and even a modest treatment is outside the scope of this text. However, it is well worthwhile to explore tutorials on this topic and to consider how different types of music support or distract from the experience. Certain types of musical forms have been employed so frequently in educational contexts that they even serve as a cultural queue to prepare us to pay attention and to convey a sense of positive expectation, technical complexity, or repetitive interactions. Tracks that embody their own transitions can also be useful when properly overlaid on transitions between different visual parts of the animation.

Technical Challenges

Software and Data Storage

The topic of data storage and related challenges of reformatting and importing are not distinct to animations but can represent an order of magnitude increase over static visualizations; aside from the source data, final high-resolution renderings, even if just a few minutes in length, can easily reach hundreds of megabytes. While such file sizes are readily manageable if you plan for it, they can at times impede interactive collaboration when colleagues are not on the same local network, such as when telecommuting.

More relevant than the sheer size of files is the challenge that might arise if a project requires migrating source data into the animation tool and there is no built-in filter for that file type. New animation developers can easily underestimate the difficulties that can arise in this step, and it is typically worthwhile to team with someone familiar with the data structures or to reach out to user groups for advice. Source data can also sometimes include unanticipated anomalies or formatting features that can require ad hoc detective work. In such situations, it might be tempting to rework data manually, viewing it as one-off effort, but in all likelihood the data will be revised and reloaded several times and so streamlining the process can be a good investment.

Scale and Time

An animation of molecular components will often use data from both higher and lower scale levels – for instance, from structural data revealing atomic structures of purified molecules, from biochemical data revealing molecular concentrations and locations, or from microscopic data – and compose them into a visually coherent whole, which can be technically challenging. Another challenge is the consistency of timescale during which one single animation is composed. The biological processes represented in an animation often occurs in different time frames, i.e. thermal motion of individual atoms goes on from 10^{-15} to 10^{-12} seconds while conformational changes and folding occurs from 10^{-6} to 10^{-1} seconds. The artist is thus often forced to remove non-critical information and compress processes. Furthermore, in order to be able to see what is actually going on, time needs to be slowed down – a lot. A useful technique can be to basically freeze-frame and enhance the contrast between the background and the key items and then point out and describe the relevant features before restarting the motion.

Dynamics

What we call macromolecules are usually flexible and dynamic entities. For example, the molecule will change its shape in response to changes in the environment, and when shape changes it is likely to influence function. When working in an animation context, it is possible to make the molecule move in two different ways: One is building the movement on simple straightforward hinge-type motions, often created by morphing between two different crystallographic states (structures of the same molecule); the other is much more complex but allows the artist to create movement closer to the real data. The second option is based on nonlinear processes using algorithms and advanced simulation approaches. No matter which option, there will almost always be a lack of kinetic information determining the movement.

Simulations seek to provide rigorous treatment of the math and science at play, where aesthetics are of lower importance. Yet that priority does not mean that some aesthetic control would not benefit. Because technical simulation software often lacks such capabilities, it can force the creator of a scientific moving picture visualization to decide between using software that can enhance aesthetics and software that provides the scientific accuracy. Animations, offering more aesthetic control in the presentation, have frequently been used to interactively present well-known concepts. While the user may already generally understand the outcome, allowing him or her to interactively alter the parameters or input information provides the opportunity to manipulate the situation and highlight a specific aspect of the physics concept.

Rigorous selectively also applies to the inclusion of interactivity and dynamics; simulations are by necessity selective, but if through seeking to be comprehensive the author embodies too many moving parts, it can easily become too confusing to illustrate and convey the core concept.

The turbulence modeling in Figure 10.4A simulates the fully developed 3D Navier-Stokes turbulent flows (providing much higher resolution and accuracy than the Navier-Stokes solver incorporated in animation software such as Autodesk Maya). Incidentally, this approach also provides a new avenue to a better and *more geometric* understanding of the problem of possible formation of singularities in the 3D Navier-Stokes model. This simulation was visualized in ParaView where the ability to make aesthetic decisions is severely limited. ParaView is an open-source multi-platform data analysis and visualization application. ParaView users can quickly build visualizations to analyze their data using qualitative and quantitative techniques. While certainly compatible with laptop usage, one of its key features is the ability to efficiently work with extremely large data sets (petascale) using distributed computing resources. On the other hand, Autodesk Maya (used for Figure 10.4B) is a general-purpose and extremely flexible tool set that readily accommodates aesthetic as well as parametric design.

In approaching these platform choices, it becomes apparent that not only are the languages of the various disciplines different but so are the types of software preferred for use in visualizations. And as mentioned earlier, one of the more complicated technical issues is exporting and importing between these different types of software.

In an attempt to unify the benefits of technical and aesthetic software, physicists are now coupling Autodesk's Maya 3D software tool with additional code to

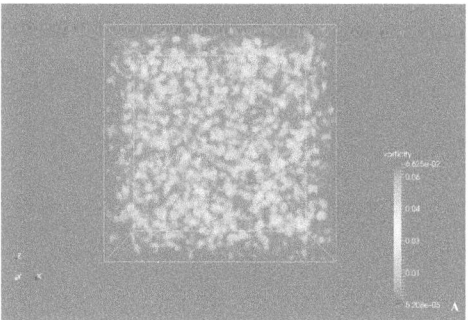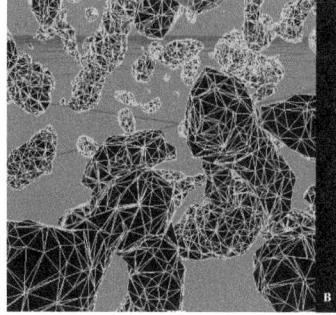

Figure 10.4 Fluid Dynamic Simulation. **A.** A still image taken from a simulation from Associate Professor Joachim Mathiesen's complex fluids research group at the Niels Bohr Institute, University of Copenhagen. The simulations visualize the interplay between the mechanisms of vortex stretching and locally anisotropic diffusion as the main physical cause of the phenomenon of turbulent dissipation. Visualized using ParaView. **B.** The same file as Figure 10.4A, viewed as a triangular mesh in Autodesk Maya versus a polymesh as seen in the ParaView visualization.

create more scientifically accurate animation and real-time simulations. In Germany, a research group coupled Maya with a two-phase Navier-Stokes fluid solver to achieve photorealistic visualization and fluid animation (Zaspel & Griebel, 2011). In Australia, a research group has designed 3D visualizations of nanostructured surfaces and bacteria using Maya. Their approach comprises a semi-automated "creative stage", where actual surface topographic parameters, obtained using an atomic force microscope, are imported into Maya via a custom Python script, and then bacterial cells and their interactions with the surfaces are visualized using available experimental data (Boshkovikj et al., 2014). These capabilities provide practical aids to knowledge discovery, assist in the dissemination of research results, and open new opportunities for engaging and educating the public.

Processes that are impossible to directly observe and also impossible to animate correctly using the most advanced animation software benefit greatly from this "combination" technique – marrying the flexibility of animation tools with the rigor of high resolution experimental data. A goal is to create visualizations that offer an almost intuitive "ah ha" grasp – to provide a stepping stone to further knowledge, elucidate abstract mathematics, and empower creative thought toward applied problems. Animation provides the tool to make science phenomena concrete, accessible, and relevant to everyday lives. The integration of simulation and animation with specific representational features is a perfect example of how these approaches directly serve to advance academic and interdisciplinary training.

While most scientists have broad experiences in viewing animations, few have formal training in how to design and create animations, outside of capturing standard output from visualization tools or adding basic enhancement to a PowerPoint or KeyNote. Like the discipline imposed by clear writing, focusing on creating a good animation can help scientists pinpoint what is important about their data. Admittedly, getting serious about such presentations can quickly lead to a sense that one is being coopted into becoming a specialist in a wholly different creative and technical domain (which is partially true). Yet there is a lot that can be done without becoming a Hollywood producer – particularly through building ongoing collaborations with other departments and specialists to create the animations. When fostering such alliances between researchers and creative professionals, shared insights are likely to be as valuable a benefit as the sheer increase in productive efficiency.

References

Bobroff, J., F. Bouquet, and C. Jutant. n.d. "New Ways to Reimagine Quantum Physics by bridging the gap between teaching and outreach". *20th International Proceedings: Conference on Multimedia in Physics Teaching and Learning.* Cited 2016 April 1. Currently under review.

Boshkovikj, V., C. J. Fluke, R. J. Crawford, and E. P. Ivanova 2014. "Three-dimensional visualization of nanostructured surfaces and bacterial attachment using Autodesk® Maya®". *Scientific Reports* 4:4228. doi:10.1038/srep04228.

Daston, L. and P. Galison. 1992. "The image of objectivity". *Representations,* 40: pp. 81–128. doi:10.2307/2928741.

Kemp, M. 2006. *Seen Unseen. Art, Science, and Intuition from Leonardo to the Hubble Telescope.* Oxford: Oxford University Press.

McGill, G. 2008. "Molecular Movies... Coming to a lecture near you". *Cell,* 133: pp. 1127–1132.

McVittie, F. 2009. "Prototypical Objects". *The Poetics of Thought*, September 18. Cited 2013 March 4. http://poeticsofthought.wordpress.com/tag/eleanor-rosch/.

Schmidt Kjærgaard, R. 2011. "Things to See and Do: How Scientific Images Work". In *Successful Science Communication. Telling it like it is*. Cambridge: Cambridge University Press.

Seiber, T. 2014. "Playable virus: HIV molecular aesthetics in science and popular culture". *Animation: An Interdisciplinary Journal* 9(2): pp. 261–76.

Wilson, E. O. et al. (2015). *Life on Earth*.

Wouters, P., F. Paas, and J. J. van Merrienboer (2008). "How to optimize learning from animated models: A review of guidelines based on cognitive load". *Review of Educational Research*, 78(3): pp. 645–675.

Zaspel, P. and M. Griebel (2011). "Photorealistic visualization and fluid animation: Coupling of Maya with a two-phase Navier-Stokes fluid solver". *Computing and Visualization in Science*, 14(8): pp. 371–383.

11 Making Sense, Nonsense, and No Sense When Representing Audio-Visual Collections

Theis Vallø Madsen

Tremendous amounts of time and resources are spent on digitizing and reorganizing collections. Often, however, collections are digitally arranged and represented in a hierarchical order where one asset or artwork is categorized as one type, from one place, with one meaning tied to one field of study or one specific institution. This way, things are categorized as from either the humanities or the natural sciences. There is nothing in between, even though digital technologies allow for less rigid categorization and representation. For instance, little effort has been put into ways of making the items in the collections accessible in other ways than traditional representation with photos and short descriptions. Collections are therefore well suited for exploratory research into new ways of arranging, navigating, and making sense in vast collections of information. Especially, audio-visual collections are interesting as case studies because of their richness of information – many artworks and audio-visual artifacts are both interesting on their own or as parts of the larger collection and its context.

The particular object and their associations force us to consider the relationship between the one and the many, the particular and the general picture. This chapter taps into broader discussions about digital culture, big data, user-generated content, and presence theory. It reconsiders methods for organizing and visualizing large data sets, in particular audio-visual collections, by addressing sense-making, nonsense-making, and no-sense-making in the work on mapping and representing these collections. Visualizing collections of art and other artifacts forces us to consider methods of sense-making and nonsense-making as a desirable byproduct of crowd-sourcing. In all this, we must not forget "no sense", or perhaps more precise "presence-making", as another byproduct of being in the world, moving about, and trying to make sense of our surroundings, including those in the digital space.

Making Sense

An unusual collection turned out to be a suitable case study for a recurring problem with large collections. The Danish artist Mogens Otto Nielsen's mail art archive at KUNSTEN Museum of Modern Art in Denmark consists of 16,000 pieces of mail art in the form of collages, postcards, letters, audiotapes, videotapes, and objects that were once circulating in the international mail art network of more than 600 artists in 42 different countries. Since the 1960s, artists have been experimenting with the building and maintaining of a decentralized and non-hierarchical network for art and information. In principle, the so-called mail art pieces were created as parts of a network, not as autonomous works of art. The network was the artwork, so to speak,

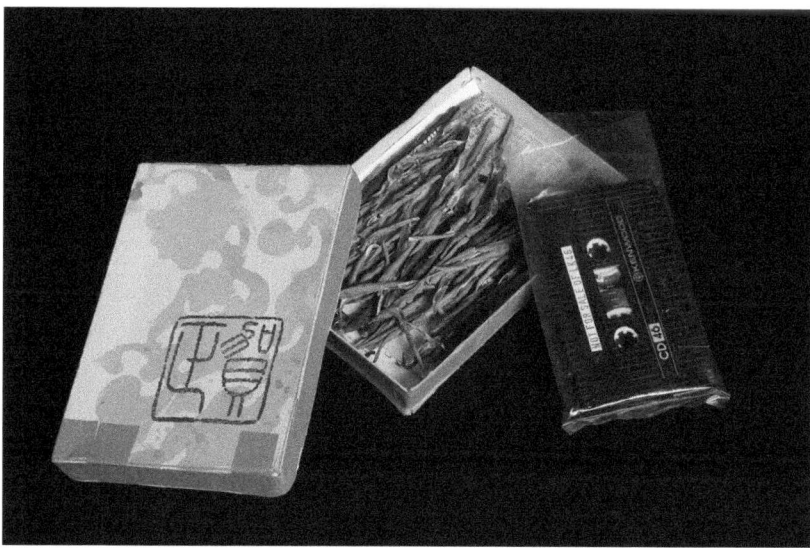

Figure 11.1 A piece in Mogens Otto Nielsen's mail art archive consists of a box containing a cassette tape and Japanese rice straws. Sent from an artist under the name of Dinn Records in Japan. Unknown date.

but a network comprised of smaller, ordinary and extraordinary pieces that were constantly modified, copied, or incorporated into new forms as they were circulating between artist hubs. One of the 16,000 pieces in Mogens Otto Nielsen's mail art archive is a cardboard box with a cassette tape, a text in Japanese, and dried Japanese rice straws (Figure 11.1).

The following accounts for the theoretical and practical considerations of the research group (of which the author was a member) that undertook the task of representing this mail art archive. Prior to research on the collection, everything was a mess in the sense that the pieces were not ordered but rather piled more or less randomly in filing cabinets and cardboard boxes with no overall registration or order. One piece may have consisted of a box that contained several other pieces, which were often related to yet other pieces. Thus, the collection required a relational database and a digital map in order to oscillate between the particular piece, groups of pieces, and the whole network.

Searching for meaning in the collection was a study into sense-making in entangled network structures. Every piece was, in principle, connected to every other piece in terms of material type, theme, artist, creative strategy, date, etc. It resembled a pre-modern cabinet of curiosities comprised of modern materials. The specific piece depicted in Figure 11.1 – a box with audio and organic materials – was part of a larger artwork with sound recordings, a book, or organic material. Making sense was a matter of analyzing bits of pieces as well as series of pieces. In order to study each piece, we would therefore have to retrace the relations between pieces as means of mapping associative meaning, movement through the network, and data analysis of, for example, the collaborative creation of pieces during the Cold War. Research was not merely a matter of art history. The postal system and this embedded artist network made it possible for artists in the West to work with artists behind the Iron Curtain during the 1970s and 1980s. The mapping of this particular collection was therefore a means of mapping Cold War history and art history as well as media history due to the fact that the artists appropriated all sorts of different media, such as faxes, cassette tapes, and videotapes along with paper, paint, and everyday materials. All sorts of materials were seemingly caught up in the artists' network and reused for artistic and other purposes. Visualizing and sense-making were therefore also a matter of mapping the objects' aesthetic, social, cultural, and political relations.

Meaning, in this case, was distributed to clusters of pieces. Collages and other artworks would often grow out of groups of artists where one facilitating artist, an artist-as-editor, would collect and redistribute pieces. When visualizing the collection, we would therefore have to retrace pieces from artist to artist in order to reconstruct each piece's history of production. This network reconstruction included artists as well as other persons, things, processes, or ideas that might have transformed a particular piece on its way to its final destination.

Mapping Mogens Otto Nielsen's mail art archive was a case study into mapping collections as dynamic networks. The method of mapping the network was based on the idea that every piece of information makes sense as part of a network. According to the actor-network theory by French sociologist Bruno Latour, everything is tied to a global and local 'network'. Pieces of information, artworks, people as well as machines and inanimate objects become local nodes in a global network (Latour 2005, 74–75). Things and ideas are never only local and never only global but always both. All things are experienced and shaped by particular circumstances such as local city planning, technical installations, and even moods of politicians and other agents in each place. In this case, the meaning of each piece in the mail art collection is distributed to artists as well as bureaucratic, technical, and institutional settings, such as postal routes and museums. The mapping of the all-embracing actor-networks would therefore include a wide range of nodes besides standard taxonomies, such as artist, country, material, and date.

Representing Networks and Meshworks

Images of networks have become common-sense models for representing complex structures. The hairy balls of gray nodes connected via ultra-thin, straight lines are supposed to represent the overall structure of any given matter. Meaning and materials travel between smaller and larger dots or nodes suspended in empty space. However, as representational models, these networks are problematic when working with collections and audio-visual archives. As British anthropologist Tim Ingold has pointed out, the nodes of networks show systems where everything of importance happens inside the nodes. They have insides and outsides, but nothing in between. Meaning or matters are somehow transported, hauled, or *beamed* from one node to the next. These common-sense images of networks represent collections as distributed or decentralized systems but also as closed-in matters separated by analytical-algebraic relations (Ingold, 2011, 63–83).

These models are suitable for visualizing electric circuits or Twitter activity, but they cannot account for heavier matters and their moving in the world. Outside electric circuits, people and their things are moving in uneven lines, in shifting pace, and in accordance to the terrain. Ingold suggests *meshworks* as a counterimage to networks (Ingold, 2011, 63). In meshworks, our things and ourselves move along lines, rarely straightforward, but usually responding and in correspondence with other lines. These lines bundle up and – unlike nodes – form knots signifying intensities or activity (Ingold, 2011, 84–102). The meshwork model portrays a world in formation. It stresses movement in favor of stationary things. If we believe that representations matter – including our images of the structure of collections – then it matters whether we visualize our networks as focal points suspended in empty space or as meshes with thicker and thinner threads. The image of meshworks reminds researchers that our objects of study are rarely transformed in secluded and self-enclosed focal points. Whereas Latour's actor-network theory allows researchers to retrace meaning in distributed network clusters, the meshwork allows us to understand meaning as something that is always in transit. In the standard network visualizations, collections are portrayed as point-to-point connections, whereas the meshwork model would, I will argue, portray the same collections as curved and crooked threads or lines. Meshworks are first and foremost a philosophical or phenomenological concept and a correction to the familiar network models rather than a method for representing collections or data. Even so, parts of the mapping of the mail art collection became a case study into mapping meshworks.

Both the network model and the meshwork model are needed in order to represent collections. The material carriers of meaning should be seen as combinations of lines and nodes, not one or the other. Network models make sense because we categorize the world as discrete items. Meshworks make sense because they show our nodes or knots as parts of lines as well as things in themselves. The knot is tied together as opposed to a dot, a spot, or a blob with insides shielded from the surroundings.

The collection of mail art pieces is a meshwork where the pieces had been reworked along the way, not merely in one or two artist hubs. Their lines of flights were important parts of their meaning and our appreciation. These movements are traceable via texts, stamps, and material in the form of worn-out envelopes, frayed edges, and grainy photocopies. For the receiving artists, the experience of unboxing and reassembling was often an important part of making sense. This opening and closing, modifying, responding, waiting, receiving, and opening again are integral to these kinds of networks or meshworks of material matters.

Mapping and making sense of the collection was twofold. On one side, the categorization and visualization should allow researchers and other users to study the particular piece, and outsiders and museum guests should be able to browse the collection. The mail art pieces were artworks, which were received and perceived as such – as precious, autonomous, and separated objects. On the other side, the categorization and visualization should allow researchers to study the relation structure – the network or meshwork – in order to study the mail art network as a historical phenomenon. Similar to the account of an aesthetic representation of the bacterium pneumococcus (Chapter 1, this volume), this re-creation of the meshwork was an attempt to create a representation that 'attracts attention [and] invokes curiosity'.

The mapping of Mogens Otto Nielsen's mail art archive became a case study for experimenting with collections of art, cultural artifacts, and information. Mapping this kind of material, researchers and digital designers had to work together in order to find a solution to the philosophical and practical problem of representing vast and entangled collections.

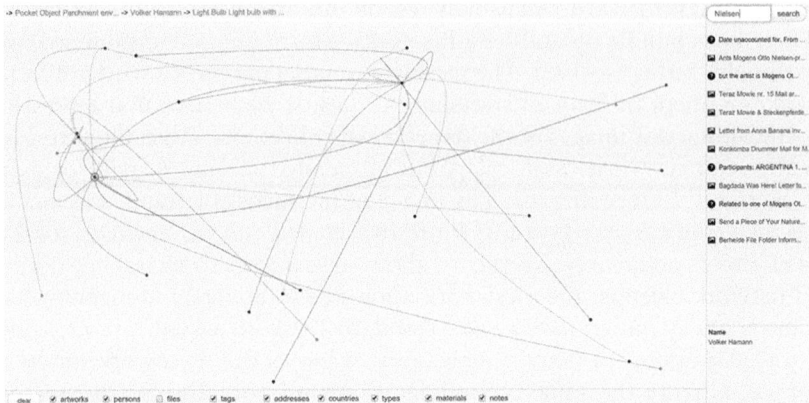

Figure 11.2 (See Color Plate 22) The first prototype of a visual online database of the mail art collection made by Martin Luckmann (interactive designer) and Theis Vallø Madsen (researcher). Each search results in the same nodes, but the relational patterns are created randomly.

Figure 11.2 (Plate 22) is a screen shot from *Prototype one* representing the collection as a crossbreed of networks and meshworks. In this visualization, the relational structure of the collection was recreated as curved, colorful lines between entry points. Continuous lines connect a specific piece to artists, other pieces, and chunks of pieces, while dotted lines are attached to tags and associations, types of material, countries, archival notes, and other taxonomies. The prototype's patterns were created automatically and at random. Researchers and other users are therefore prevented from taking the beaten path. They are forced to follow the new threads of the unpredictable patterns. The prototype, then, sacrificed a bit of order to represent or recreate randomness as an integral part of searching and consulting these kinds of collections.

Making Nonsense

The mail art network is largely unknown in both art communities and the broader public. One of the reasons is the simple fact that the historical mail art network had no audience (Madsen, 2015, 6). From the beginning in the early 1950s, the mail art network was an attempt to circumvent the traditional supply line of art going from the atelier of a genius artist to galleries via art buyers or museum professionals. In the mail art network, one had to be a node in the network in order to receive and experience the artworks. Perceiving was hinged on receiving. In as much, there were no masterpieces, only an abundance of mail art pieces and related material.

Nonsense was an essential part of the network. The ongoing exchange and the expansion of the network were more important than the individual artworks that were exchanged. The pieces were often created with inspiration from the Dadaists in the 1910s using nonsense as an aesthetic and political strategy in the critique of established institutions. In the mail art network, nonsense was encouraged by the use of secret codes and regulations that were undecipherable for an unintended audience outside the network. In addition, one had to study every piece in the specific exchange. Nonsense, then, was part of an aesthetic strategy as well as an effect of the structure of the network.

In recent years, these accumulations of crowd build and entangled mail art collections are being relocated to museums from artists' attics, basements, and spare rooms. When we catalogue, digitalize, and visualize these collections, we must account for this history of production. Visualizing a mail art collection is an attempt to visualize the work of a crowd rather than a collection of traditional or modern art. The clutter cannot and should not be edited out entirely. Nonsense was a tactical part of building an artists' network with other principles than the artworks created for an audience in a gallery or a museum. Nonsense, then, should be a part of the representation of the collection in order to show the nature of the artists' network.

Recreating the crowd-built collection meant that smaller and larger crowds should once again access, annotate, and – to a certain degree – even edit mail art collections. It was never meant to be a static, complete collection. Instead, it was part of its *raison d'être* to expand and facilitate change. Evidence of this attitude is a repeated rubber-stamped statement like one that appears on several of Mogens Otto Nielsen's send-outs from the 1980s and onward declaring that 'ALL REPRODUCTION • MODIFICATION • DERIVATION AND TRANSFORMATION OF THIS OBJECT IS PERMITTED'.

In that sense, artists' network, in principle, was working as a peer-to-peer network and an open-source community before those terms became fashionable (Madsen, 2015, 87–88). Similar to three of the five principles of digital information, as described by media theorist Lev Manovich, the material was *variable*, *modular*, and dependent on *transcoding* (Manovich, 2001, 27–48; Madsen, 2015, 54–56). Artworks were reworked again and again by different artists, thereby turning them into variables, and artworks were used as building blocks in new pieces, thereby making them modular. Transcoding – the exchange between artists – was both a means and an end in the production of these pieces in an attempt to create artworks that would take a life on its own inside the network. The mapping project would therefore have to include time, movement, and transformation in the core of the visualization – not as a neat add-on feature.

Crowds can be difficult to contain and control. Yet, as demonstrated by Chapter 8 (this volume) and its account of citizen science, crowds can work more efficiently and more intelligently than the authoritative sources. The benefits of *swarm intelligence* or the so-called wisdom of the crowd has been known since British scientist Francis Galton studied the phenomenon in the beginning of the 20th century. After visiting a fair, Galton discovered that a crowd of about 800 people could guess the actual weight of a bull with a margin of 1 pound (Galton, 1907, 450–451). In vast collections and other big data sets, crowd-sourcing can contribute with information in terms of metadata and relations. In the specific case of the mail art collection, the sheer quantity and its piecemeal character stress the need for users to annotate and associate pieces. There is too much information, too much material, and too many relations for one or two researchers to retrace and reconstruct a usable network of the collection. This collection was in dire need of outsiders to enrich its metadata – much like *Galaxy Zoo* where volunteers assess astronomical images and help scientists distinguish spiral galaxies from elliptic ones (Chapter 8, this volume).

The determining factor is whether or not the data set has the proper tools and user interface for access and editing. The frustrations and fascinations with the fleeting nature of Nielsen's mail art were the offset of a research group consisting of art historians, archivists, and biologists with collections of mail art, historical photographs, and preserved butterflies, birds, and other animals. In *Mapping the Archive project,* these researchers joined with digital designers and dramaturges in order to develop new methods for using, expanding, and navigating in audio-visual archives and collections.

The collections of mail art, historical photographs, and dead birds (and other samples of the local fauna) were rich in terms of information. Despite their differences, all three collections needed a digital visual map that would enable users to annotate, sort out, and link materials. There was a need for easy, visible navigation in the huge data sets – whether those were the specific route of a mail art piece, the geographical location of an old photograph, or the migratory routes of a specific group of birds. The preliminary results are described below as *Prototype two* (Figure 11.3, Plate 23), but the work is ongoing and, probably, never ending. Thread-making is re-interpreted in this second prototype as small, shuffling bacteria in a digital Petri dish.

The second prototype attempted to visualize the vibrating nature of the collection. The visualization re-imagined each piece as a small, shuffling bacteria in a digital Petri dish. Relations between the vibrant pieces were a matter of distance and size instead of point-to-point connections. In the online version of the prototype, users would see

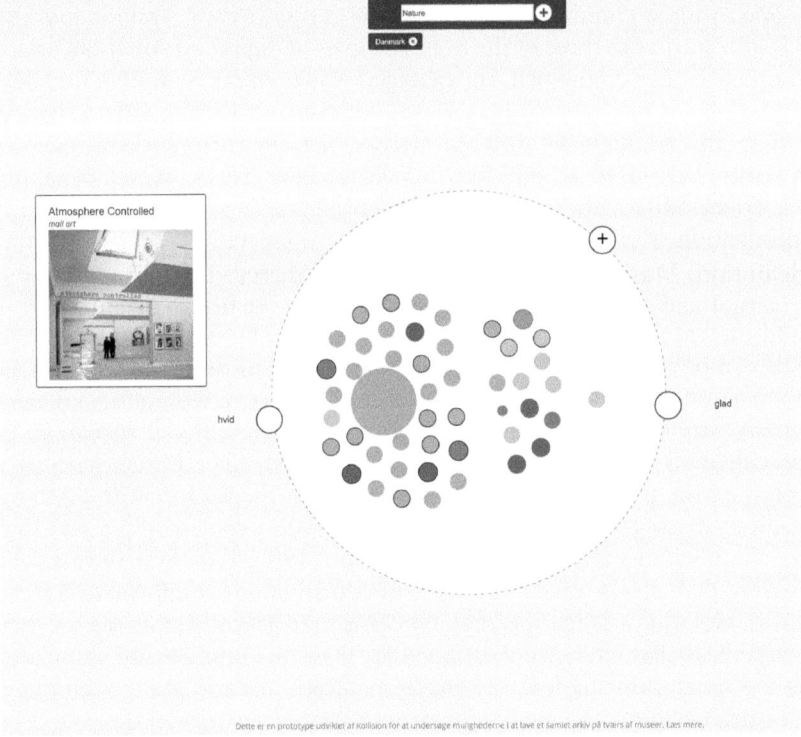

Figure 11.3 (See Color Plate 23) The second prototype (Danish version) for a visual online database of the mail art collection. Items are attracted to certain categories. Colored dots (mail art pieces) are suspended between a magnet tool set on "hvid" (white) and a magnet tool set on "glad" (happy). Searching for "Nature" will pull out orange dots (specimens from the Museum of Natural History in Aarhus). Users can group items and add tags. Made by Kollision Aps and *The Mapping the Archive* project.

a small selection of the collection depending on searches. Size of cells would increase and decrease when they used specific tags.

In this representation, the images and items from the collections were not *in* a category but related to categories. In the outer rim of the digital Petri dish, users were able to visually attract certain pieces. The tool behaved like adjustable magnets that could be set to attract pieces of a certain color or emotional tag. Users could, for instance, use one magnet tool to pull pieces containing the color "white" while another magnet tool allowed users to pull or pieces that were labeled "happy". When a piece would be equally attracted to both the yellow-set magnet tool and mellow-set magnet tool, the piece would then be suspended in the space between the two. Each piece was not, then, in either one or the other but suspended between categories. These digitally represented pulls and repulsions were an attempt to work around the habit of visualizing and categorizing our collections as fluid instead of fixed. The items in this particular collection cannot be contained in one category, for example either private or public, performative or object-based, network or extraordinary, but applies to more than one category. They are in between the categories that are usually used to describe and place the content of collections.

When visualizing a collection, the initial mappings and categorizations create the framework for further additions. Once a certain order has settled, our use and understanding of the objects will usually follow the order, either by continuation or negation. Organizing principles prefer certain connections, series, and hierarchies in collections, and they hide as much as they reveal (Foucault, 1972, 129). Even if the metadata of the mail art collection is crowd built, the researcher and the structures of the museum database create its basic order and arrangement. Future user-generated content is therefore inevitably tied to the choices of the researcher and the system.

Representing the collection using photographs and scans have a stronger appeal to users than symbols and signs, as Lev Manovich points out in "What is Visualization?". In his definition, Manovich argues for the use of 'direct visualization' comprised of 'real' or actual audio-visual material, when possible, instead of graphical primitives (Manovich, 2011, 36). However, unlike Manovich, I will argue that visualizations are always primitive or reductive regardless of their size and complexity. One is always representing when arranging, and one is always framing when visualizing. With regards to museum collections, visualizations and other representational methods have a certain obligation to present themselves as *re*presentations, i.e. interpretations, first and foremost, because the pieces in these collections are richer than the photos of them and, second, because these institutions must allow others to study and criticize their methods for producing and representing knowledge and history.

This obligation should also be extended to work on arranging and visualizing audio-visual collections. Extending a collection into digital space is also a matter of curating the collection, leading users to certain nodes, texts, or images while neglecting others. Navigating between piles of materials, the challenge is to clear a pathway for users leading them toward the most important pieces while also allowing less-disciplined users to go astray and discovers areas outside the well-lit spots in the collection.

In most cases, we tend to go with the flow – whether the flow would lead via curated pathways or the ways of the crowd. In the first prototype – although never implemented – the curated routes and popular user-generated routes would become stronger or denser in accordance with the traffic. The user-generated pathways that were rarely used would gradually fade out and disappear. The visualization would not be a static image but something that swells and shrinks during the course of time.

In *Mapping the Archive*, we worked with color tagging as well as tentative emotional tagging. While color mapping can be done automatically, the idea of emotional mapping is more questionable. Emotions are circumstantial, individual, and unquantifiable. Yet emotions are part of these collections and should not be ignored as a waste product of knowledge production.

Likewise, there are indeed plenty of practical, philosophical, and scientific reasons to be critical toward ideas of quantifying something as momentary as art and emotions. Yet these attempts might reveal or confirm that certain things cannot be exchanged and numbered. In this particular case, the experiment created new points of entry to the specific collections. It was not a choice between sense and emotion, objectivity and subjectivity, or the certain and the uncertain but a way to allow for a wider range of perspectives on pieces and the overall collection. These experimental approaches were never meant to substitute standard taxonomies or orders but merely to point to alternative and creative ways of mapping collections with digital tools.

Making No Sense

When we talk about meaning-making and sense in vast collections, we should be equally concerned with the aesthetics of presence of such collections. In *Production of Presence* from 2004, the German-American literary theorist Hans Ulrich Gumbrecht describes presence as a 'state of being lost in focused intensity' (Gumbrecht, 2004, 104). The experience of presence is fleeting and unpredictable but nevertheless an essential part of what attract us to music, museums, literature, archives, or artworks. They create moments of intensity, which is different from the work of interpretation, that is searching for 'meaning' as structures of signs and a set of choices that have been taken as opposed to alternative choices (Gumbrecht, 2004, 105).To consider presence is therefore to account for all those effects that escape our language or meaning in the traditional sense (Gumbrecht, 2004, 124). 'Sense-making' in presence theory is tied to the sensory apparatus, not a well-hidden and predetermined *meaning* either below or behind the surface. In this sense, presence theory is a method for creating representations of collections that shape and support anticipation, curiosity, and wonder.

Things have concrete and sensational qualities alongside those pieces of information that make sense and meaning in the traditional sense of the word. *Presence effects* are often left out when we describe, categorize, and visualize, although the particular atmosphere and presence are usually things we remember from encounters with art, artifacts, events, or assets in collections. When in a museum, we rarely pay attention to the floors, its material feels, and our choreographed walk through passages and along walls. Instead, we might talk about atmosphere or a specific piece we stumbled upon unexpectedly. The aesthetic of presence is connected to aura, atmosphere, affect, and other attempts to capture and describe the experience and effect of space and material.

Writing about atmosphere, German philosopher Gernot Böhme distinguished between two kinds of real, *Realität* and *Wirklichkeit*. The former includes hard facts, colors, age, materials, and all those things, which are collectively known as factual. The latter, our *Wirklichkeit* is determined by our own conception of the real – our ascription of meaning to *Realität*. It is what we do or make out of it (Böhme, 2003; Bjerregaard, 2014, 4). Working on ways to visualize and represent collections, we should account for both kinds of real. Representing vast collections should enable researchers and others to make sense, nonsense, and no sense. The three modes of working with collections should supplement one another.

Digital technologies are often portrayed as antithetical to authenticity and true, deep, or real aesthetical experience. As Linda Stone, former executive at Apple and Microsoft, wrote in an online magazine:

> Before the Internet, I made more trips to the library and more phone calls. I read more books and my point of view was narrower and less informed. I walked more, biked more, hiked more, and played more. I made love more often.
>
> (Stone, 2011, 218)

The digital space is considered a poorer space without atmosphere and potential for authentic experiences. The idea of past technologies as realer or truer has a long history of philosophers and others describing modernity as a transformation of reality into simulacra or of loss of aura (Baudrillard, 2007, 1018–1019; Benjamin, 1980, 475; Madsen, 2011, 80–81). In this schema, the dichotomy places *presence* in 'real' three-dimensional space, in opposition to an assumed superficiality or inauthenticity of virtual space.

Western visual culture has been deeply concerned with representing three-dimensional space on a two-dimensional plane since, at least, the Renaissance and its new and scientific methods for representing space using central perspective and grids (Madsen, 2011, 80–81). A contemporary image of this striving for more accurate representations of space can be seen when visiting virtual museums. In these new digital museums, we can observe a special kind of data representation. In recent years, cultural institutions have invited people on virtual tours and into virtual spaces with virtual collections. Museums have virtual models where visitors can explore the panoramic view of the gallery space (see, for example, British Museum on Google Museum View). Instead of walking, we are beamed or hauled from point to point in the same way that Google street view hauls us forward from vantage point to vantage point while hovering above the pavement. In other virtual representations of spaces, the museum space is rebuilt as a clean, ceilingless cube in empty space.

Naturally, these representations are merely meant as a model of the real. However, the virtual models and maps misrepresent the spaces and collections as well. They repeat an old idea about art and collections as things for the bodiless, noise-free mind. On the virtual tours, we float around in space as drones or disembodied eyes. These visualizations simulate a visit to the museum, but they neglect to inform its users of the loss of touch, atmosphere, material, light, sounds, smell, the importance of moving along the walls, across the floors, entering rooms from a certain angle, bending down, etc.

Mining archives and collections are embodied ways of navigating as well. We search by numbers and tags but also by weight of a file folder, the size of cases, vague ideas of location in paper piles as well as numerous possible associations and circumstances from earlier visits. Reading, weighing, tracing, listening, fumbling, stumbling from document to document are common-sense ways of searching in large quantities of material (Madsen, 2015, 171). In digital user interfaces, we scan surfaces, touch screens, or move a mouse. We click and swipe. These smaller corrections and adjustments are well suited for searching in data. However, they cannot replicate a body and a mind moving through galleries or buildings. Poor imitations of three-dimensional space are therefore a wrong turn when we want to use digital technologies to represent and visualize collections.

Digital technology is not *per se* unsuitable for representing collections, even though we should abandon the idea of virtual reconstructions. Visualizations and mapping can be a tool for the production of sense, nonsense as well as presence. Digital tools are useful for making sense in big data sets, but we need to figure out ways to balance between the search for meaning and a sense of presence. Digital maps of collections might be excellent tools for leading users and researchers toward effects and experiences. The challenge is therefore not to create a replica but to use the digital space on its own terms. Its maps should direct our attention to its blind spots, not ignore them. The abovementioned work, prototyping a digital map to vast collections were unsatisfying in finding a one-size solution to the problem. Yet the work enabled researchers to reconsider ways to access, navigate, associate, and make sense while refraining from making sense where there was none.

References

Baudrillard, J. 2007. "The Hyper-realism of Simulation". In *Art in Theory 1900–2000*, edited by C. Harrison and P. Wood. Oxford: Blackwell Publication, 1992/2007 (1976).

Benjamin, W. 1980. *Das Kunstwerk im Zeitalter seiner technischen Reproduzierbarkeit*, third edition. Frankfurt: Suhrkamp, 1980 (1935).

Böhme, G. 2003. "Space as Bodily Presence and Space as Medium for Representations". In *Transforming Spaces: The Topological Turn in Technology Studies*, edited by M. Hård, A. Lösch, and D. Verdicchio, Dirk. Cited 2015 November 16.

Bjerregaard, P. 2014. "Dissolving objects: Museums, atmosphere and the creation of presence". *Emotion, Space and Society*. doi: 10.1016/j.emospa.2014.05.002.

Foucault, M. 1972. *The Archaeology of Knowledge—and the Discourse on Language*, translated by A. M. Sheridan Smith. New York: Pantheon Books, 1972 (1971).

Galton, F. 1907. "Vox Populi". *Nature* 1949(75): pp. 450–451. Cited 2015 October 15. http://galton.org/essays/1900-1911/galton-1907-vox-populi.pdf.

Gumbrecht, H. U. 2004. *Production of Presence: What Meaning Cannot Convey*. Cambridge and London: Harvard University Press, 2004.

Ingold, T. 2011. *Being Alive: Essays on Movement, Knowledge and Description*. New York: Routledge, 2011.

Latour, B. 2005. *Reassembling the Social: An Introduction to Actor-Network-Theory*. Oxford and New York: Oxford University Press, 2005.

Madsen, T. V. 2011. *Genfortryllelse i den digitale tidsalder*. Aarhus, Denmark: Digital Aesthetics Research Center, 2011.

Madsen, T. V. 2015. "Ants in the Archive. Cataloguing Mogens Otto Nielsen's Mail Art Archive". PhD diss., Aarhus University, Denmark.

Manovich, L. 2001. *The Language of New Media*. Cambridge and London: The MIT Press, 2001.

Manovich, L. 2011. "What is visualization?" *Visual Studies* 26(1): pp. 36–49.

Stone, L. 2011. "Navigating Physical and Virtual Lives". In *How is the Internet Changing the Way You Think?*, edited by J. Brockman. London: Atlantic Books.

12 "Facts" – and Representational Acts

Morten Kyndrup

The previous chapters in this volume have demonstrated the complexity of the problems concerning illustrations of scientific work and representations of scientific data. A variety of questions have been raised, especially about the status and function of visual representations in comparison with the scientific "truth content" they are supposed to present. What is the relationship between the aesthetic quality of a representation and the clarity/truth of its scientific substance – complementary, parallel, potentially clashing (Chapters 2 and 9 of this volume)? Should the "form" of a representation, at all levels, be dictated by the intended "function"; should the "how" depend on the "what" exclusively (Chapter 3)? May salience and relevance be mismatched directly (Chapter 3)? Are basically different modes of reception connected to "aesthetic" and "scientific" perception respectively – and how are they interconnected, if even so (Chapters 4 and 7)? If it is true that only form with a function is beautiful (Chapter 3), where does that leave "art" and its alleged autonomy? What are actually so-called "raw data", which our pretty pictures are supposed to represent? Are these "data" actually taking part in both the phenomenon and its representation, and thereby in fact being anything but "raw" (Chapter 6)? What is the subsequent function of representations (and of representational conventions) – do they actually influence the development direction of science itself actively, thus possessing an epistemic function themselves by being both products and producers of knowledge (Chapter 6)? Are there specific epistemic consequences of the widespread use of storytelling and animations in representations to varying audiences – given the fact that (implicit) narrators necessarily possess a distinctive authority (Chapters 8 and 10)?

Questions like these have been raised and discussed through the chapters of this volume. The present chapter does not pretend to be able to deliver proper solutions to these problems. It does, however, propose an approach, a perspective, by means of which to clarify and mutually contextualize these questions, hopefully making them easier both to handle, practically, and to further develop, theoretically.

The approach of this chapter is to see representations, not as finalized entities, but as *acts* – representational acts of communication. Acting implies agency, and as a point of departure, this chapter introduces a model of communication, including these (embedded and "real") instances of acting. In this chapter, I argue that although the (representational) act as such is one, it may possess different kinds and levels of functions, varying modes of addressing, all in all, a cluster of communicational implications and potentials, corresponding to similar distinctions within and between its model and actual recipients. This chapter covers the problem of coding, framing, and/or contextualizing and defines and discusses the question of "aesthetic" relations or dimensions and their representational functions. Following this definition and discussion, I argue that a specific negotiation of the role of each kind of function, within the representational act in question, is necessary – and sketch out the conditions for such negotiation. The chapter also touches on the distinctive problems about possible epistemic consequences of the representations themselves, in that connection, and concludes by proposing a distinctive division of labor in the production of representations of scientific data.

Signification as an Act

As Émile Benveniste (1966) convincingly argued in his revision of Ferdinand de Saussure's basic work in semiotics, signs, in themselves, cannot produce meaning through their sheer, conventional connection to their proper "signified". Meaning; semantics, is engendered only in concrete situations by significational acts, including specific agents. The sentence, "We are here now", exclusively consists of shifters, each of them without any detached, clear reference. Consequently, the sentence makes no "sense" as such, in itself. But a sentence like that makes perfect sense when uttered by someone in a situation by which the "we", the "are", the "here", and the "now" are all well defined – that is, without a situation, without utterance or enunciation, no signification, no "meaning". Since any utterance presupposes agents – someone to speak (write, draw, sing, etc.), and someone to listen (read, see, etc.) – meaning is always produced by way of acts.

This fact is condensed in Karl Bühler's classical model of communication, showing the four basic instances involved in any significational act (Figure 12.1A): a "sender" (Se) and a "receiver" (Rc), a "sign" (Sg) and a "referent" Rf). Someone says something (by means of signs, referring to something else) to somebody.

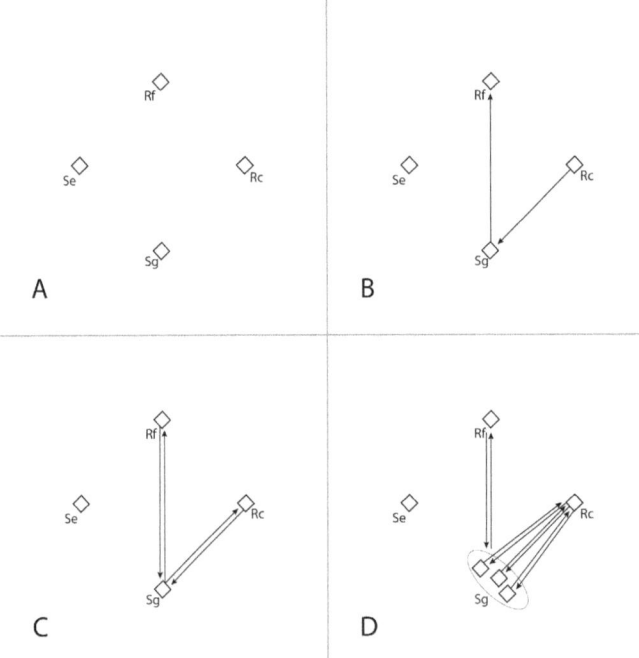

Figure 12.1 Karl Bühler's classical model of communication redrawn.

Now, when in a specific situation we read a figure that is presented as a representation of scientific data, say, for instance Figure 6.1 (Plate 13) from Chapter 6, we actually participate in a real communication act. We try to understand what the colors, the arrow, and the limit between two halves of the picture stand for, i.e. "mean", what they represent of conducted research, and which documented "truth" they incarnate. We will be conscious of the fact that someone has figured out these exact colors, limits, and the arrow in order to clarify to us exactly this specific "truth" about the abilities of different methods of microscopy. And we will probably be aware that the "sign" or "data representation" we perceive refers to "data", which, in themselves, are always already mediating between phenomenon and representation, as thoroughly discussed in Chapter 6.

But apart from just reading a figure in a real act of enunciation while actually being its receiver, I may approach this figure analytically. As it turns out, any written (drawn, reproduced) significational act actually includes its fundamental communicational instances in terms of traces of what is called embedded or implied instances or positions (this fact is technically called "enunciated enunciation"; see Greimas & Courtès, 1988, 279f; Kyndrup 2003). These implied positions might be analyzed and described by their individual properties and through the specific dynamic, internal relations between the embedded positions of the representation in question. Let us for instance regard the implied Sg↔Rc relation in Figure 6.1 (Plate 13). The embedded receiver is obviously presupposed to possess specific competences as a precondition to partake in the communication offered by the representation. These competences span from low-key abilities, such as being able to distinguish red from green, to decoding the arrow as a meta-level pointing out a specific intention of the implied sender – and to more advanced skills, such as knowledge about advanced biology and the methodological history of microscopy. If we were regarding a piece of representation in the shape of an animation, one might concordantly point out which level of movie perception competence the supposed receiver should specifically possess in order to be able to understand the animation in question. But one might, of course, also take a closer look at the specific Sg↔Rf relation in this case. How does this figure represent the "data" in question? Which modifications have been made, and what are their consequences (does salience match relevance)? Are the differentiations, for instance in terms of choice of colors, actually adequately reproducing the facts found in the research? Is the relation between the representation of scientific data and the "raw data" true in the sense that we are dealing with a realistic depiction "of" something – or is the representation rather symbolically representing something actually not visible (as in the case of Figure 5.5, Plate 12, "displaying" black matter proved by an absence, i.e. elliptically represented through "gravitational lensing")? Similarly, the implied sender position may be analyzed and localized in its intentions, competences, possible errors, and shortcuts the way these occur from the representation itself (including of course supplementary information from paratexts, such as headings, legends, etc.).

Significational Functions

Although a given representation of scientific data indeed constitutes one, and only one, representational act, it may have multiple significational functions, and it may emphasize and prioritize these functions differently through the construction of its embedded positions and relations.

As a point of departure, any representation of scientific data is made for cognitive purposes. It aims at making the result of a piece of scientific work (some kind of "raw data") accessible and understandable to somebody (else). It should literally be able to guide its receiver through the representation (Sg) to the represented (Rf) so that the scientific "truth" of Rf, including its evidence in the research conducted, is shared convincingly by the receiver (see Figure 12.1B), i.e. is becoming *her* truth as well.

The basis for this is of course the assumption that "raw data" are usually in themselves not capable of creating this effect of understanding, acceptance, and conviction (see the discussion in Chapter 3 on "raw data" versus possible interpretations). They need "help" in the shape of a representation, e.g. in the shape of a visualization, which can enhance the understanding by offering an interpretation of the data.

In the process of producing a representation, main attention will be directed toward two internal relations: The Rc↔Sg and the Sg↔Rf (Figure 12.1C). Concerning the Rc↔Sg relation, whoever elaborates the representation must make sure that its sign-form is accessible to the intended receiver. Its (visual) language must be understandable and unambiguous. In order to ensure that, a clear analysis of the specifically required properties and abilities of the model receiver must be made. This, in turn, calls for clear choices of the audience aimed for (spanning from experts in the field, exclusively, to lay people), and it presupposes knowledge about average "reading skills" in different population groups. On top of that, detailed knowledge about specific meaning effects of varying visual expression types is required. Different colors have varying, specific effects concerning quantification, and graphs, for instance, may be difficult to understand depending on one's educational background.

But the Sg↔Rf relation as well calls for careful attention during the process. It must be ensured that the representation (Sg) is adequate in the sense that it is still a "true" depiction of the data results. And what does "true" mean in this relationship? If different colors are used for showing a distinction found in the material, is the representation then "true", if these colors are in fact not matching the actual "colors" of the object(s), or if the object(s) have no colors (cf. once again Figure 5.5, Plate 12)? This of course has to be considered carefully. By all means, a translation into another media is taking place, the purpose of which is to make a piece of scientific knowledge accessible. This wish for accessibility often makes some kind of adaptation or simplification necessary. The consideration will be, then, to which degree you can simplify/clarify while still keeping intact the truth content concerned.

The above reflections deal with ideal typical cases exclusively. In real-life practice, things are much more complex, and representational interests may be constituted by very diverse and even contradictory components. What if, for some reason, you would like to make a representation overtly attractive and pleasant? Or challenging in a specific way? Or directly entertaining? Or even "beautiful" in its own right? We will come back to the issue of "beauty". By all means, the constitution of a representation (sign) in these cases would have to be influenced by considerations, which were not natural parts of the original cognitive purpose. None of such "mixed" considerations, it should be underlined, are necessarily disturbing, restrictive, or irrelevant to the fundamental cognitive purpose of the representation. But they may cause considerable bias, and they should be carefully analyzed by whatever consequences they might have.

Notwithstanding the composition of desired and realized clusters of functions in a specific representation, it is important to maintain the fact that we are still dealing with one singular act of signification. One act, multiple functions.

Framing: Codes, Contexts, and Conventions

In reality, of course, no act of communication takes place on a deserted island, as a free interplay between imagined positions, as in the Bühler-model above. All signifiers, both in verbal and in visual language, are linked to signified meanings through conventional codes. These codes were historically created, and we have learned them as part of our cultural encyclopedia. This "we", however, has different levels, and it must be differentiated when analyzing both production and reception of significational acts. The "we" of Western civilization in general, that of Danish white, middle-class intellectuals and that of professional visual artists in New York are not one and the same "we", and thus the encoding of capacities and rules differ. Furthermore, besides sociologically defined contexts, more local contexts will play a part in deciding the meaning of a given act. This volume may serve as an example of such specific context. Given the subject of this volume, all illustrations in the chapters are to be understood, not primarily as cognitive tools to enhance perception of scientific results (which, for most of them, were their original purpose), but in this context rather at a meta-level as illustrations of possible ways of depicting such results. No act of signification can be understood without taking its specific context into consideration. This "context", however, always consists of a complex system of levels and layers, some engendered historically and others defined locally, all of them partaking in the resulting meaning of a given act. As expressed in a pun by the literary theorist Jonathan Culler: 'Meaning is context-bound, and context is boundless' (Culler, 1997, 67). You can never exhaustingly describe each and every contextual influence on a given significational act; you will have to settle for just calculating the possible effects of the most decisive ones when analyzing specific expressions.

This context question is important to the representation of scientific data discussion here, at several levels. One is of course the fact, also pointed out in several of the previous chapters, that specific conventions for representation are being developed, some of them just locally within distinctive disciplines, others more in general. Molecules tend to be depicted in specific ways in distinctive disciplines; again and again, we meet certain colors symbolizing temperature distinctions and/or pollution levels. The development of such "local" conventions for representations are advantageous in the sense that receivers learn to decode illustrations more immediately and thus to understand them better and more easily. But any convention also implies a constraint: You cannot just ignore an existing convention when depicting this or that; you will need to relate to what is already given in the field – and this may mean a delimitation of the potentials for an adequate or clarifying illustration of a given result. A good example of this is given in the report, in Chapter 6 of this volume, explaining how, in a specific case, some physicists had to change a depiction into an "older" and less adequate conventional version in order to be sure that biologists could make use of it.

Far more important in the representation of scientific data, however, is the question of framing, in terms of deciding the specific *type* of functional context – and the significational consequences of this decision or classification. As a point of departure, any representation of scientific data belongs to a strictly scientific context, serving a cognitive purpose exclusively. It is meant to facilitate and to enhance the understanding of a specific "truth" about something, defined within the framework of science, i.e. obeying the rules and complying with the conventions for scientific knowledge. The adequacy and the purpose of the representation are primarily and neatly controlled by transparent rules and expectations along the Sg↔Rf and the Sg↔Rc axes, respectively. As a receiver, you know that you are supposed to *learn* from this, to acquire knowledge, which in turn may be shared with others, subsequently. The "clean" representation is, metaphorically spoken, just a device for facilitating the transport of a specific piece of knowledge.

But what happens when representations of scientific data become more "mixed" or ambiguous in their functions and, say, start signaling that they are (also) pieces of entertainment? As a receiver, you may still, to some extent, expect to learn, but in a context of entertainment, you would probably also expect to have fun, be engaged, "kill time", forget yourself... Your set of criteria for evaluating the representation will have changed, and concordantly, the representation would obviously have been produced differently. If we go one step further and think of a representation contextualizing itself as "beautiful", or even as belonging to the realm of "art", a further blurring or differentiation of the meaning of this representation would take place. The purpose and function of the representation would no longer be "clean" and unambiguous. The implication of this is not that a representation could not be perfectly cognitively helpful *and* highly entertaining *and* strikingly beautiful at the same time – just that a contextual and thus meaning-engendering complexity like that would have to be specifically dealt with. Dealt with by receivers, of course, but not least dealt with by producers.

Aesthetic Value

Generally, the very concept of "aesthetics" and the predicate "aesthetic" are unfortunately used in a highly imprecise way, both in everyday language and in scholarly contexts. This (uncoincidental) fact is not least due to the history of the concept, including the occurrence of certain competing and almost contradictory traditions from Romanticism and onward. But these ambiguities are also connected to the extremely complex character of the concept itself (for a thorough elaboration of these problems, see Kyndrup, 2008). "Aesthetics" as a word and a concept was not invented until the mid-18th century and was born as the designation of a new discipline dealing with a specific kind of cognition. Coined as such by Alexander Gottlieb Baumgarten, but substantially developed just a few years later by Immanuel Kant (1968), the concept of "aesthetic value" is defined as something that is engendered through a specific kind of *relation* between us and things, subjects and objects. An aesthetic relation includes a judgment, in which "I" evaluate the properties of a given object as it appears "for me", but conceiving and pronouncing this judgment as if it were referring to an objective concept of beauty, shared by everyone, all the while I know that such a general, shared concept of aesthetic quality does not exist. That is, aesthetic judgments are always concrete, situation-bound, and singular, but they are pronounced and shared as if they were universal. Aesthetic judgments as such thus constitute the creation of "passages" between the "I/me", the world (of objects in one object), and the (notion of a) "we". But these passages are engendered only through a distinct mode of perception, in which we evaluate something as itself, the way it appears and has value for me. This is not about pointing out anything else about this "thing" as a referent or a sign (standing for something true, right, sensible, evil, etc.). It is about what the thing is in and by itself, the way it appears to "me" in this specific evaluating relation. It is thus not the object that makes the relation aesthetic; it is the relation that makes the object aesthetic (Genette, 1997, 18). This distinctive aesthetic way of connecting ourselves to things is particularly brought in function in our approach to artworks, not least due to historically important parallels between aesthetic relations and the constitution of the autonomy of art and thus its "purposelessness". But aesthetic qualities are not to be found exclusively in our relation to art and artworks. We do establish aesthetic relations and thus ascribe aesthetic value also to many other things, from objects of design to landscapes or sunsets.

Utterly important in connection with the discussion of representations of scientific data is the fact that "beauty", aesthetic value, inherently has nothing to do with "truth" or any other kind of "transitive" purpose. Beauty, as conceived in Modernity, simply has no extrinsic purpose; it is there for me as a momentum of my perception of exactly this object (which of course does not prevent the existence and development of conventions for beauty). But basically beauty constitutes a purpose in itself, by and in the very way in which it exists – not by what it might refer to or stand for. Actually, something may be beautiful *and* false, evil, inappropriate, or useless. This distinction is one of those, which only Modernity makes historically possible.

Negotiation

From the above definition, it is obvious that a representation of scientific data, intrusively inviting to be perceived as an aesthetic object, will constitute a kind of contradiction to its own fundamental purpose. On the one hand, the Sg position is supposed to refer to some scientific results/"raw data" (Rf) in such a way that understanding and acknowledgement is enhanced and facilitated. This signifying act thus has a clear cognitive purpose, and the piece of knowledge intended to be passed on is, per definition, transferable in its capacity of its status as "scientific", complying with the institutionalized conventions for that specific kind of truth. On the other hand, "beauty", aesthetic value of the Sg position seen as an independent entity, as a thing in itself, i.e. perceived explicitly *not* as something primarily with a representing, referring function, will be extraneous to the general purpose of the representation. So, following this logic, one should carefully avoid too beautiful illustrations in the case of representations of scientific data in order not to derail the perception and the attention of the receivers.

It is, however, not substantiated that the interrelationship between significational functions should be complementary, making an "aesthetic" perception of a given sign/form automatically an acknowledgment of, say, cognitive functions of the same sign prohibitive or at least difficult. It is not like in the sharing-a-cake model, where "more for me" means "less for you"; this either/or is a myth (although widespread). The coexistence of detached significational functions, within one and the same signifying act, does indeed constitute an utterly complex communication situation, in which differing appeal structures, contextualization, and implied intentions are mixed together. The good news is, however, that our general competences of reception have developed concordantly and that these competences actually match this complexity. All of us can easily distinguish between, say, the "beauty" of our new iPad and its value for us as users, or between that of our brand-new, smart high-heeled shoes and the practical problems about wearing them throughout an entire party and getting sores and blisters. We are perfectly used to negotiating and handling different kinds of "value" and to dealing with detached levels of signifying functions in the "same" objects or situations. All kinds of communication are mixed, are composite, and are in a way messed up in insurmountable complexities of mutually incongruent levels and functions. Still, however, we are used to dealing with these complexities. We are trained not just to navigate within and to respond to these complexities, but we are actually capable of reaching concrete conclusions when surrounded by them. We know we need to choose between, say beauty, user value, the cultural signals, and the price to pay when considering to buy or to do or to perceive something. And if we can combine different layers in order to optimize the meeting of our actual needs, we will do it.

The acknowledgment of this complexity of functions, and of our acquired competences to deal with them practically, does not, however, imply that "anything goes" in all communicational acts, neither that choices, priorities, and distinctions should not have specific irreversible consequences. In all cases, including those of representations of scientific data, the answer is negotiation. But, by all means, negotiation on the basis of an open recognition of the constitutional incompatibility of the multiple, signifying functions in play. This means that one can never boil down the variety of potential functions in any given act to just one resulting one (which some might dream of in the case of representations of scientific data). But this also means that multi-functionality is not necessarily a contrast to clarity.

A representation of scientific data of obvious high aesthetic quality, i.e. in itself calling for aesthetic appreciation as "beautiful", may very well enhance the attractiveness of the Sg position as such and may thus also be helpful to the basic function of this Sg position: its reference to some scientific data. This "help" may be found at various levels, one of them being the receiver's sheer appreciation of being addressed aesthetically, of being asked to judge even at this level, in a situation originally dealing with something completely different (transmission of scientific information). It should, however, be stressed that although these functions may be "helpful" to each other, they cannot be unified and ranked along the same axis. Some seem to believe that mere form is pretty, but form with function is beautiful. Such a statement, although intuitively understandable, tends to reduce the detachedness, the distinctiveness of the functional levels respectively, into one resultant. Something exquisitely "beautiful" at the Sg position in a representation may also very well lead to a closure of the transparency of the position and thus make the representation opaque and dysfunctional, where its cognitive purposes are concerned. But no general recipes, no rules about how to add "more" or "less" aesthetic quality exist. We are dealing with multiple, mutually incongruent functions within one representational act, and the specific distribution of levels, their construction and functions, respectively, will have to be calculated to be negotiated carefully by the producers of a representation in each and every case. All choices made will have their consequences, their price to be paid or to be rewarded in terms of their specific impact on the signifying act in question. At the end of the day, any such act is singular and should be produced, analyzed, and received as such.

Division of Labor

As it is, the complexity of representations of scientific data in general has, to an increasing degree, been recognized also within science itself. This recognition implies that the admission that presentation, illustration, and mediation of scientific results are far too complicated to leave to amateurs. They are tasks for professionals. But who are actually the professionals in this connection? The scientists behind the results themselves? The artists/craftsmen who produce the actual pictures/animations/representations? The communication analysts who consider the effects of the different layers and functions within the representation in question? The publishers of journals and magazines, the curators at the museums of science, etc.?

The answer is that no single profession covers this entire complex field, and that none of the above fields of expertise can be omitted. Although, in the end, any representation of scientific data constitutes one singular act of signification, this act should appear as the result of professional insights and skills of very different kinds. The scientists are necessary in order to ensure the truth of the reference function back to the actual results. Artists/craftsmen/technicians are necessary in order to produce a representation optimizing its possibilities in terms of distinctions, accessibility, and aesthetic challenges. Analysts are necessary in order to calculate the potential specific effects of the choices made and thus to decide how to combine the detached parts or functions of the act. Publishers, curators – in short, those responsible for the channels of intermediation – are of course necessary in order to ensure that the representations of scientific data are properly fitted into their respective channels. On top of all of this, these different approaches and pieces of work must necessarily be coordinated and processed by someone.

All in all, a carefully calculated, coordinated, and realized division of labor is required (as demonstrated in Chapters 9 and 11 in this volume). This is no big surprise, since division of labor is the core principle behind all civilizational development and thus the basis for our accumulating knowledge and wealth. Still, however, today there is a reluctance toward accepting this division of labor in connection with representations of scientific data in particular.

This reluctance is neither surprising nor irrational, especially not due to its historical background. The commitment of science to a referential, documented truth has created an understandable skepticism toward any kind of simplification, modification, or beautification in the representations – artists, analysts, and publishers basically being suspected of following other interests than the precise presentation of a scientific truth content.

What is needed, however, is a division of labor based on mutual confidence – not just someone hiring different experts to fulfill defined tasks, thus splitting up the process. Division, in principle, implies that each and every participant, responsible for a detached part of the process, also joins the overall responsibility for the act as a whole. The adequacy of any detail forms part of the functions of the act as a whole, and these functions actually require a variety of competences to be understood properly.

This call for division is, of course, substantiated by the ambition of obtaining the best results in representations of scientific data. Best results are important, not least because we know that these results may form the basis also for further scientific progress and/or will create conventions for future representations in the field (see for instance the important discussion of the "epistemic functions" of presentations in Chapter 6 of this volume).

But division, including shared full responsibility, is also a precondition for avoiding some of the obvious dangers at hand. One example of such a danger might be the so-called 'edutainment' business. As described previously (in Chapter 8 of this volume), science and technology promoting to the masses may also be a big-money business.

Billions are spent to build new science museums, and these institutions may not aim primarily at presenting scientific truths but rather at making money. This means that, for instance, the criterion of optimizing entertainment may easily be given priority to that of maintaining scientific truth. In these "sugar-coating" processes, we may end up with presentations containing next to nothing but sugar, i.e. primarily aimed at offering smooth, pleasant, and highly entertaining experiences, able to compete with any other bargain in our experience society. But of course profit-making is not the only possible threat to an optimized balance in representations of scientific data. An exaggerated aesthetic elaboration of a presentation may be problematic if, in the end, it monopolizes all attention and thereby actually blocks the representational function of the Sg position. Again, division of labor with joint responsibility is the best way of avoiding that kind of bias as well.

Presenting science to the masses, making big business out of science presentation, creating beautiful presentations, even to the point of becoming genuine artworks, using exquisitely seductive presentation modes such as storytelling and animation – none of these tasks necessarily constitute a problem per se. But it is important to keep in mind what this is actually all about: It is about science, scientific results, proven, documented, sharable knowledge about the world we live in. At the end of the day, science has one, and only one, pertinence criterion, and that is its "truth".

But in conclusion, nothing necessarily becomes less true just by being entertaining, engaging, attractive, or even beautiful in addition to being true. By regarding and analyzing representations of scientific data in their status as signifying acts with several detached functions, these functions may be calculated and negotiated in each their own status and role in a given Sg↔Rc relation and, besides that, may be, individually and together, measured against how they influence the Sg↔Rc relation, the pertinence and adequacy of which the whole operation of course is all about (Figure 12.1D).

That being said, it should be noticed that referential cognition and aesthetic appreciation, in fact, have not little in common, when it comes to the mode and type of perception processes. At issue in both cases is a specific individual in the process of recognizing something and, by doing that, simultaneously experiencing herself as the "I", which is here and now acquiring knowledge and/or judging aesthetical qualities. In both processes one sees oneself see, both processes are reflexive. On the basis of that kinship, why not dream of these processes being further developed into supporting and strengthening each other when forming parts of the same signifying act?

References

Baumgarten, A. G. 1988 [1750/58]. *Theoretische Ästhetik. Die grundlegenden Abschnitte aus der "Aesthetica"*, translated and edited by H. R. Schweizer. Hamburg: Felix Meiner.

Benveniste, É. 1966. *Problèmes de linguistique générale I-II*. Paris: Gallimard.

Culler, J. 1997. *Literary Theory: A Very Short Introduction*. Oxford: Oxford University Press.

Genette, G. 1997. *L'oeuvre de l'art **. La relation esthétique*. Paris: Seuil.

Greimas, A. J. and J. Courtès. 1988. *Semiotik. Sprogteoretisk ordbog*, edited by P. A. Brandt and O. Davidsen. Aarhus, Denmark: Aarhus University Press.

Kant, I. 1968 [1790]. *Kritik der Urteilskraft -Werkausgabe Band X*, edited by W. Weischedel. Frankfurt am Main: Suhrkamp.

Kyndrup, M. 2003. *Kunstværk og udsigelse (ACTS 18)*. Aarhus, Denmark: Aarhus University.

Kyndrup, M. 2008. *Den æstetiske relation. Sanseoplevelsen mellem kunst, videnskab og filosofi*. København: Gyldendal.

Contributors

Ebbe Sloth Andersen Associate Professor at Interdisciplinary Nanoscience Center and Department of Molecular Biology and Genetics, Aarhus University, Denmark.

Ditte Høyer Engholm PhD Stipend at Department of Molecular Biology and Genetics, Aarhus University, Denmark.

David S. Goodsell Associate Professor of Molecular Biology at Department of Integrative Structural and Computational Biology, The Scripps Research Institute, California, and Research Professor at Center for Integrative Proteomics Research, Rutgers, the State University of New Jersey.

Steen Hannestad Professor at Department of Physics and Astronomy, Aarhus University, Denmark.

Mogens Kilian Professor Emeritus in Medical Microbiology, Professor at Institute of Biomedicine – Research and Education, Aarhus University, and Adjunct Professor in Bacterial Population Genetics, University of Copenhagen, Denmark.

Rikke Schmidt Kjærgaard Jens Chr. Skou Junior Fellow at Aarhus Institute of Advanced Studies, Aarhus University, and Associate Professor at Interdisciplinary Nanoscience Center – INANO-MBG, Aarhus University, Denmark.

Kelly Krause Creative Director, Nature, Nature Research Group, London.

Martin Krzywinski Staff Scientist at Canada's Michael Smith Genome Sciences Center at the BC Cancer Research Center, Vancouver.

Morten Kyndrup Professor in Aesthetics and Culture, and Director at Aarhus Institute of Advanced Studies, Aarhus University, Denmark.

Theis Vallø Madsen Postdoctoral Fellow at Faaborg Museum and Institute of Cultural Science, University of Southern Denmark, Denmark.

Bjørn Panyella Pedersen AIAS-COFUND (Marie Curie) Junior Fellow at Aarhus Institute of Advanced Studies, Aarhus University, and Assistant Professor at Institute of Molecular Biology and Genetics, Aarhus University, Denmark.

Lotte Philipsen Associate Professor at School of Communication and Culture, Art History, Aarhus University, Denmark.

Janet Rafner Fulbright Fellow at Aarhus University and Niels Bohr Institute, University of Copenhagen, Denmark.

Nina Samuel Art & Science Historian, Postdoctoral Researcher at Image, Knowledge, Gestaltung: An Interdisciplinary Laboratory, Cluster of Excellence at Humboldt University, Berlin and at Zentrum für Literatur- und Kulturforschung, Berlin.

Morten Søndergaard Associate Professor at Institute of Communication, Aalborg University, Denmark.

Frederik Stjernfelt Professor at Institute of Communication, Aalborg University, Denmark.

Anette Vandsø Postdoctoral Fellow at ARoS Art Museum and School of Communication and Culture, Aarhus University, Denmark.

Djuke Veldhuis AIAS-COFUND (Marie Curie) Junior Fellow at Aarhus Institute of Advanced Studies, Aarhus University, Denmark.

Index